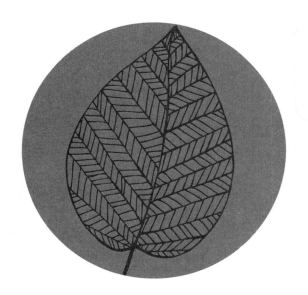

Patterns and the Plant World

Grade 1

STEM Road Map
for Elementary School

Edited by Carla C. Johnson, Janet B. Walton, and
Erin Peters-Burton

National Science Teachers Association

Arlington, Virginia

National Science Teachers Association

Claire Reinburg, Director
Rachel Ledbetter, Managing Editor
Deborah Siegel, Associate Editor
Andrea Silen, Associate Editor
Donna Yudkin, Book Acquisitions Manager

ART AND DESIGN
Will Thomas Jr., Director, cover and
    interior design
Himabindu Bichali, Graphic Designer, interior
    design

PRINTING AND PRODUCTION
Catherine Lorrain, Director

NATIONAL SCIENCE TEACHERS ASSOCIATION
David L. Evans, Executive Director

1840 Wilson Blvd., Arlington, VA 22201
www.nsta.org/store
For customer service inquiries, please call 800-277-5300.

FSC
www.fsc.org
MIX
Paper from
responsible sources
FSC® C011935

NSTA is committed to publishing material that promotes the best in inquiry-based science education. However, conditions of actual use may vary, and the safety procedures and practices described in this book are intended to serve only as a guide. Additional precautionary measures may be required. NSTA and the authors do not warrant or represent that the procedures and practices in this book meet any safety code or standard of federal, state, or local regulations. NSTA and the authors disclaim any liability for personal injury or damage to property arising out of or relating to the use of this book, including any of the recommendations, instructions, or materials contained therein.

Library of Congress Cataloging-in-Publication Data
Names: Johnson, Carla C., 1969- editor. | Walton, Janet B., 1968- editor. | Peters-Burton, Erin E., editor.
Title: Patterns and the plant world, grade 1 : STEM road map for elementary school /
    edited by Carla C. Johnson, Janet B. Walton, and Erin Peters-Burton.
Description: Arlington, VA : National Science Teachers Association, [2018] | Includes bibliographical references
    and index.
Identifiers: LCCN 2018019327 (print) | LCCN 2018023480 (ebook) | ISBN 9781681405087 (e-book) |
    ISBN 9781681405070 (print)
Subjects: LCSH: Seasons--Study and teaching (Elementary) | Plants--Seasonal variations--Study and teaching
    (Elementary)
Classification: LCC QB637.2 (ebook) | LCC QB637.2 .P38 2018 (print) | DDC 508.2--dc23
LC record available at *https://lccn.loc.gov/2018019327*

The *Next Generation Science Standards* ("NGSS") were developed by twenty-six states, in collaboration with the National Research Council, the National Science Teachers Association and the American Association for the Advancement of Science in a process managed by Achieve, Inc. For more information go to *www.nextgenscience.org*.

CONTENTS

CONTENTS

ABOUT THE EDITORS AND AUTHORS

Dr. Carla C. Johnson is the associate dean for research, engagement, and global partnerships and a professor of science education at Purdue University's College of Education in West Lafayette, Indiana. Dr. Johnson serves as the director of research and evaluation for the Department of Defense–funded Army Educational Outreach Program (AEOP), a global portfolio of STEM education programs, competitions, and apprenticeships. She has been a leader in STEM education for the past decade, serving as the director of STEM Centers, editor of the *School Science and Mathematics* journal, and lead researcher for the evaluation of Tennessee's Race to the Top–funded STEM portfolio. Dr. Johnson has published over 100 articles, books, book chapters, and curriculum books focused on STEM education. She is a former science and social studies teacher and was the recipient of the 2013 Outstanding Science Teacher Educator of the Year award from the Association for Science Teacher Education (ASTE), the 2012 Award for Excellence in Integrating Science and Mathematics from the School Science and Mathematics Association (SSMA), the 2014 award for best paper on Implications of Research for Educational Practice from ASTE, and the 2006 Outstanding Early Career Scholar Award from SSMA. Her research focuses on STEM education policy implementation, effective science teaching, and integrated STEM approaches.

Dr. Janet B. Walton is a research assistant professor and the assistant director of evaluation for AEOP at Purdue University's College of Education. Formerly the STEM workforce program manager for Virginia's Region 2000 and founding director of the Future Focus Foundation, a nonprofit organization dedicated to enhancing the quality of STEM education in the region, she merges her economic development and education backgrounds to develop K–12 curricular materials that integrate real-life issues with sound cross-curricular content. Her research focuses on collaboration between schools and community stakeholders for STEM education and problem- and project-based learning pedagogies. With this research agenda, she works to forge productive relationships between K–12 schools and local business and community stakeholders to bring contextual STEM experiences into the classroom and provide students and educators with innovative resources and curricular materials.

Dr. Erin Peters-Burton is the Donna R. and David E. Sterling endowed professor in science education at George Mason University in Fairfax, Virginia. She uses her experiences from 15 years as an engineer and secondary science, engineering, and mathematics teacher to develop research projects that directly inform classroom practice in science and engineering. Her research agenda is based on the idea that all students should build self-awareness of how they learn science and engineering. She works to help students see themselves as "science-minded" and help teachers create classrooms that support student skills to develop scientific knowledge. To accomplish this, she pursues research projects that investigate ways that students and teachers can use self-regulated learning theory in science and engineering, as well as how inclusive STEM schools can help students succeed. During her tenure as a secondary teacher, she had a National Board Certification in Early Adolescent Science and was an Albert Einstein Distinguished Educator Fellow for NASA. As a researcher, Dr. Peters-Burton has published over 100 articles, books, book chapters, and curriculum books focused on STEM education and educational psychology. She received the Outstanding Science Teacher Educator of the Year award from ASTE in 2016 and a Teacher of Distinction Award and a Scholarly Achievement Award from George Mason University in 2012, and in 2010 she was named University Science Educator of the Year by the Virginia Association of Science Teachers.

Dr. Andrea R. Milner is the vice president and dean of academic affairs and an associate professor in the Teacher Education Department at Adrian College in Adrian, Michigan. A former early childhood and elementary teacher, Dr. Milner researches the effects constructivist classroom contextual factors have on student motivation and learning strategy use.

Dr. Tamara J. Moore is an associate professor of engineering education in the College of Engineering at Purdue University. Dr. Moore's research focuses on defining STEM integration through the use of engineering as the connection and investigating its power for student learning.

Dr. Vanessa B. Morrison is an associate professor in the Teacher Education Department at Adrian College. She is a former early childhood teacher and reading and language arts specialist whose research is focused on learning and teaching within a transdisciplinary framework.

Dr. Toni A. Sondergeld is an associate professor of assessment, research, and statistics in the School of Education at Drexel University in Philadelphia. Dr. Sondergeld's research concentrates on assessment and evaluation in education, with a focus on K–12 STEM.

ACKNOWLEDGMENTS

This module was developed as a part of the STEM Road Map project (Carla C. Johnson, principal investigator). The Purdue University College of Education, General Motors, and other sources provided funding for this project.

See *www.routledge.com/products/9781138804234* for more information about *STEM Road Map: A Framework for Integrated STEM Education.*

PART 1

THE STEM ROAD MAP

BACKGROUND, THEORY, AND PRACTICE

OVERVIEW OF THE *STEM ROAD MAP CURRICULUM SERIES*

Carla C. Johnson, Erin Peters-Burton, and Tamara J. Moore

The *STEM Road Map Curriculum Series* was conceptualized and developed by a team of STEM educators from across the United States in response to a growing need to infuse real-world learning contexts, delivered through authentic problem-solving pedagogy, into K–12 classrooms. The curriculum series is grounded in integrated STEM, which focuses on the integration of the STEM disciplines—science, technology, engineering, and mathematics—delivered across content areas, incorporating the Framework for 21st Century Learning along with grade-level-appropriate academic standards.

The curriculum series begins in kindergarten, with a five-week instructional sequence that introduces students to the STEM themes and gives them grade-level-appropriate topics and real-world challenges or problems to solve. The series uses project-based and problem-based learning, presenting students with the problem or challenge during the first lesson, and then teaching them science, social studies, English language arts, mathematics, and other content, as they apply what they learn to the challenge or problem at hand.

Authentic assessment and differentiation are embedded throughout the modules. Each *STEM Road Map Curriculum Series* module has a lead discipline, which may be science, social studies, English language arts, or mathematics. All disciplines are integrated into each module, along with ties to engineering. Another key component is the use of STEM Research Notebooks to allow students to track their own learning progress. The modules are designed with a scaffolded approach, with increasingly complex concepts and skills introduced as students progress through grade levels.

The developers of this work view the curriculum as a resource that is intended to be used either as a whole or in part to meet the needs of districts, schools, and teachers who are implementing an integrated STEM approach. A variety of implementation formats are possible, from using one stand-alone module at a given grade level to using all five modules to provide 25 weeks of instruction. Also, within each grade band (K–2, 3–5, 6–8, 9–12), the modules can be sequenced in various ways to suit specific needs.

STANDARDS-BASED APPROACH

The *STEM Road Map Curriculum Series* is anchored in the *Next Generation Science Standards (NGSS)*, the *Common Core State Standards for Mathematics (CCSS Mathematics)*, the *Common Core State Standards for English Language Arts (CCSS ELA)*, and the Framework for 21st Century Learning. Each module includes a detailed curriculum map that incorporates the associated standards from the particular area correlated to lesson plans. The STEM Road Map has very clear and strong connections to these academic standards, and each of the grade-level topics was derived from the mapping of the standards to ensure alignment among topics, challenges or problems, and the required academic standards for students. Therefore, the curriculum series takes a standards-based approach and is designed to provide authentic contexts for application of required knowledge and skills.

THEMES IN THE *STEM ROAD MAP CURRICULUM SERIES*

The K–12 STEM Road Map is organized around five real-world STEM themes that were generated through an examination of the big ideas and challenges for society included in STEM standards and those that are persistent dilemmas for current and future generations:

- Cause and Effect
- Innovation and Progress
- The Represented World
- Sustainable Systems
- Optimizing the Human Experience

These themes are designed as springboards for launching students into an exploration of real-world learning situated within big ideas. Most important, the five STEM Road Map themes serve as a framework for scaffolding STEM learning across the K–12 continuum.

The themes are distributed across the STEM disciplines so that they represent the big ideas in science (Cause and Effect; Sustainable Systems), technology (Innovation and Progress; Optimizing the Human Experience), engineering (Innovation and Progress; Sustainable Systems; Optimizing the Human Experience), and mathematics (The Represented World), as well as concepts and challenges in social studies and 21st century skills that are also excellent contexts for learning in English language arts. The process of developing themes began with the clustering of the *NGSS* performance expectations and the National Academy of Engineering's grand challenges for engineering, which led to the development of the challenge in each module and connections of the module activities to the *CCSS Mathematics* and *CCSS ELA* standards. We performed these mapping processes with large teams of experts and found that these five themes provided breadth, depth, and coherence to frame a high-quality STEM learning experience from kindergarten through 12th grade.

Cause and Effect

The concept of cause and effect is a powerful and pervasive notion in the STEM fields. It is the foundation of understanding how and why things happen as they do. Humans spend considerable effort and resources trying to understand the causes and effects of natural and designed phenomena to gain better control over events and the environment and to be prepared to react appropriately. Equipped with the knowledge of a specific cause-and-effect relationship, we can lead better lives or contribute to the community by altering the cause, leading to a different effect. For example, if a person recognizes that irresponsible energy consumption leads to global climate change, that person can act to remedy his or her contribution to the situation. Although cause and effect is a core idea in the STEM fields, it can actually be difficult to determine. Students should be capable of understanding not only when evidence points to cause and effect but also when evidence points to relationships but not direct causality. The major goal of education is to foster students to be empowered, analytic thinkers, capable of thinking through complex processes to make important decisions. Understanding causality, as well as when it cannot be determined, will help students become better consumers, global citizens, and community members.

Innovation and Progress

One of the most important factors in determining whether humans will have a positive future is innovation. Innovation is the driving force behind progress, which helps create possibilities that did not exist before. Innovation and progress are creative entities, but in the STEM fields, they are anchored by evidence and logic, and they use established concepts to move the STEM fields forward. In creating something new, students must consider what is already known in the STEM fields and apply this knowledge appropriately. When we innovate, we create value that was not there previously and create new conditions and possibilities for even more innovations. Students should consider how their innovations might affect progress and use their STEM thinking to change current human burdens to benefits. For example, if we develop more efficient cars that use by-products from another manufacturing industry, such as food processing, then we have used waste productively and reduced the need for the waste to be hauled away, an indirect benefit of the innovation.

The Represented World

When we communicate about the world we live in, how the world works, and how we can meet the needs of humans, sometimes we can use the actual phenomena to explain a concept. Sometimes, however, the concept is too big, too slow, too small, too fast, or too complex for us to explain using the actual phenomena, and we must use a representation or a model to help communicate the important features. We need representations and models such as graphs, tables, mathematical expressions, and diagrams because it makes our thinking visible. For example, when examining geologic time, we cannot actually observe the passage of such large chunks of time, so we create a timeline or a model that uses a proportional scale to visually illustrate how much time has passed for different eras. Another example may be something too complex for students at a particular grade level, such as explaining the *p* subshell orbitals of electrons to fifth graders. Instead, we use the Bohr model, which more closely represents the orbiting of planets and is accessible to fifth graders.

When we create models, they are helpful because they point out the most important features of a phenomenon. We also create representations of the world with mathematical functions, which help us change parameters to suit the situation. Creating representations of a phenomenon engages students because they are able to identify the important features of that phenomenon and communicate them directly. But because models are estimates of a phenomenon, they leave out some of the details, so it is important for students to evaluate their usefulness as well as their shortcomings.

Sustainable Systems

From an engineering perspective, the term *system* refers to the use of "concepts of component need, component interaction, systems interaction, and feedback. The interaction of subcomponents to produce a functional system is a common lens used by all engineering disciplines for understanding, analysis, and design." (Koehler, Bloom, and Binns 2013, p. 8). Systems can be either open (e.g., an ecosystem) or closed (e.g., a car battery). Ideally, a system should be sustainable, able to maintain equilibrium without much energy from outside the structure. Looking at a garden, we see flowers blooming, weeds sprouting, insects buzzing, and various forms of life living within its boundaries. This is an example of an ecosystem, a collection of living organisms that survive together, functioning as a system. The interaction of the organisms within the system and the influences of the environment (e.g., water, sunlight) can maintain the system for a period of time, thus demonstrating its ability to endure. Sustainability is a desirable feature of a system because it allows for existence of the entity in the long term.

In the STEM Road Map project, we identified different standards that we consider to be oriented toward systems that students should know and understand in the K–12 setting. These include ecosystems, the rock cycle, Earth processes (such as erosion, tectonics, ocean currents, weather phenomena), Earth-Sun-Moon cycles, heat transfer, and the interaction among the geosphere, biosphere, hydrosphere, and atmosphere. Students and teachers should understand that we live in a world of systems that are not independent of each other, but rather are intrinsically linked such that a disruption in one part of a system will have reverberating effects on other parts of the system.

Optimizing the Human Experience

Science, technology, engineering, and mathematics as disciplines have the capacity to continuously improve the ways humans live, interact, and find meaning in the world, thus working to optimize the human experience. This idea has two components: being more suited to our environment and being more fully human. For example, the progression of STEM ideas can help humans create solutions to complex problems, such as improving ways to access water sources, designing energy sources with minimal impact on our environment, developing new ways of communication and expression, and building efficient shelters. STEM ideas can also provide access to the secrets and wonders of nature. Learning in STEM requires students to think logically and systematically, which is a way of knowing the world that is markedly different from knowing the world as an artist. When students can employ various ways of knowing and understand when it is appropriate to use a different way of knowing or integrate ways of knowing, they are fully experiencing the best of what it is to be human. The problem-based learning scenarios provided in the STEM Road Map help students develop ways of thinking like STEM professionals as they ask questions and design solutions. They learn to optimize the human experience by innovating improvements in the designed world in which they live.

THE NEED FOR AN INTEGRATED STEM APPROACH

At a basic level, STEM stands for science, technology, engineering, and mathematics. Over the past decade, however, STEM has evolved to have a much broader scope and broader implications. Now, educators and policy makers refer to STEM as not only a concentrated area for investing in the future of the United States and other nations but also as a domain and mechanism for educational reform.

The good intentions of the recent decade-plus of focus on accountability and increased testing has resulted in significant decreases not only in instructional time for teaching science and social studies but also in the flexibility of teachers to promote authentic, problem solving–focused classroom environments. The shift has had a detrimental impact on student acquisition of vitally important skills, which many refer to as 21st century skills, and often the ability of students to "think." Further, schooling has become increasingly siloed into compartments of mathematics, science, English language arts, and social studies, lacking any of the connections that are overwhelmingly present in

the real world around children. Students have experienced school as content provided in boxes that must be memorized, devoid of any real-world context, and often have little understanding of why they are learning these things.

STEM-focused projects, curriculum, activities, and schools have emerged as a means to address these challenges. However, most of these efforts have continued to focus on the individual STEM disciplines (predominantly science and engineering) through more STEM classes and after-school programs in a "STEM enhanced" approach (Breiner et al. 2012). But in traditional and STEM enhanced approaches, there is little to no focus on other disciplines that are integral to the context of STEM in the real world. Integrated STEM education, on the other hand, infuses the learning of important STEM content and concepts with a much-needed emphasis on 21st century skills and a problem- and project-based pedagogy that more closely mirrors the real-world setting for society's challenges. It incorporates social studies, English language arts, and the arts as pivotal and necessary (Johnson 2013; Rennie, Venville, and Wallace 2012; Roehrig et al. 2012).

FRAMEWORK FOR STEM INTEGRATION IN THE CLASSROOM

The *STEM Road Map Curriculum Series* is grounded in the Framework for STEM Integration in the Classroom as conceptualized by Moore, Guzey, and Brown (2014) and Moore et al. (2014). The framework has six elements, described in the context of how they are used in the *STEM Road Map Curriculum Series* as follows:

1. The STEM Road Map contexts are meaningful to students and provide motivation to engage with the content. Together, these allow students to have different ways to enter into the challenge.

2. The STEM Road Map modules include engineering design that allows students to design technologies (i.e., products that are part of the designed world) for a compelling purpose.

3. The STEM Road Map modules provide students with the opportunities to learn from failure and redesign based on the lessons learned.

4. The STEM Road Map modules include standards-based disciplinary content as the learning objectives.

5. The STEM Road Map modules include student-centered pedagogies that allow students to grapple with the content, tie their ideas to the context, and learn to think for themselves as they deepen their conceptual knowledge.

6. The STEM Road Map modules emphasize 21st century skills and, in particular, highlight communication and teamwork.

All of the STEM Road Map modules incorporate these six elements; however, the level of emphasis on each of these elements varies based on the challenge or problem in each module.

THE NEED FOR THE *STEM ROAD MAP CURRICULUM SERIES*

As focus is increasing on integrated STEM, and additional schools and programs decide to move their curriculum and instruction in this direction, there is a need for high-quality, research-based curriculum designed with integrated STEM at the core. Several good resources are available to help teachers infuse engineering or more STEM enhanced approaches, but no curriculum exists that spans K–12 with an integrated STEM focus. The next chapter provides detailed information about the specific pedagogy, instructional strategies, and learning theory on which the *STEM Road Map Curriculum Series* is grounded.

REFERENCES

Breiner, J., M. Harkness, C. C. Johnson, and C. Koehler. 2012. What is STEM? A discussion about conceptions of STEM in education and partnerships. *School Science and Mathematics* 112 (1): 3–11.

Johnson, C. C. 2013. Conceptualizing integrated STEM education: Editorial. *School Science and Mathematics* 113 (8): 367–368.

Koehler, C. M., M. A. Bloom, and I. C. Binns. 2013. Lights, camera, action: Developing a methodology to document mainstream films' portrayal of nature of science and scientific inquiry. *Electronic Journal of Science Education* 17 (2).

Moore, T. J., S. S. Guzey, and A. Brown. 2014. Greenhouse design to increase habitable land: An engineering unit. *Science Scope* 51–57.

Moore, T. J., M. S. Stohlmann, H.-H. Wang, K. M. Tank, A. W. Glancy, and G. H. Roehrig. 2014. Implementation and integration of engineering in K–12 STEM education. In *Engineering in pre-college settings: Synthesizing research, policy, and practices*, ed. S. Purzer, J. Strobel, and M. Cardella, 35–60. West Lafayette, IN: Purdue Press.

Rennie, L., G. Venville, and J. Wallace. 2012. *Integrating science, technology, engineering, and mathematics: Issues, reflections, and ways forward.* New York: Routledge.

Roehrig, G. H., T. J. Moore, H. H. Wang, and M. S. Park. 2012. Is adding the *E* enough? Investigating the impact of K–12 engineering standards on the implementation of STEM integration. *School Science and Mathematics* 112 (1): 31–44.

STRATEGIES USED IN THE
STEM ROAD MAP CURRICULUM SERIES

Erin Peters-Burton, Carla C. Johnson, Toni A. Sondergeld, and Tamara J. Moore

The *STEM Road Map Curriculum Series* uses what has been identified through research as best-practice pedagogy, including embedded formative assessment strategies throughout each module. This chapter briefly describes the key strategies that are employed in the series.

PROJECT- AND PROBLEM-BASED LEARNING

Each module in the *STEM Road Map Curriculum Series* uses either project-based learning or problem-based learning to drive the instruction. Project-based learning begins with a driving question to guide student teams in addressing a contextualized local or community problem or issue. The outcome of project-based instruction is a product that is conceptualized, designed, and tested through a series of scaffolded learning experiences (Blumenfeld et al. 1991; Krajcik and Blumenfeld 2006). Problem-based learning is often grounded in a fictitious scenario, challenge, or problem (Barell 2006; Lambros 2004). On the first day of instruction within the unit, student teams are provided with the context of the problem. Teams work through a series of activities and use open-ended research to develop their potential solution to the problem or challenge, which need not be a tangible product (Johnson 2003).

ENGINEERING DESIGN PROCESS

The *STEM Road Map Curriculum Series* uses engineering design as a way to facilitate integrated STEM within the modules. The engineering design process (EDP) is depicted in Figure 2.1 (p. 10). It highlights two major aspects of engineering design—problem scoping and solution generation—and six specific components of working toward a design: define the problem, learn about the problem, plan a solution, try the solution, test the solution, decide whether the solution is good enough. It also shows that communication

Figure 2.1. Engineering Design Process

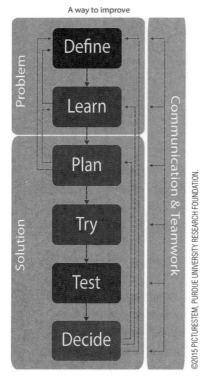

and teamwork are involved throughout the entire process. As the arrows in the figure indicate, the order in which the components of engineering design are addressed depends on what becomes needed as designers progress through the EDP. Designers must communicate and work in teams throughout the process. The EDP is iterative, meaning that components of the process can be repeated as needed until the design is good enough to present to the client as a potential solution to the problem.

Problem scoping is the process of gathering and analyzing information to deeply understand the engineering design problem. It includes defining the problem and learning about the problem. Defining the problem includes identifying the problem, the client, and the end user of the design. The client is the person (or people) who hired the designers to do the work, and the end user is the person (or people) who will use the final design. The designers must also identify the criteria and the constraints of the problem. The criteria are the things the client wants from the solution, and the constraints are the things that limit the possible solutions. The designers must spend significant time learning about the problem, which can include activities such as the following:

- Reading informational texts and researching about relevant concepts or contexts

- Identifying and learning about needed mathematical and scientific skills, knowledge, and tools

- Learning about things done previously to solve similar problems

- Experimenting with possible materials that could be used in the design

Problem scoping also allows designers to consider how to measure the success of the design in addressing specific criteria and staying within the constraints over multiple iterations of solution generation.

Solution generation includes planning a solution, trying the solution, testing the solution, and deciding whether the solution is good enough. Planning the solution includes generating many design ideas that both address the criteria and meet the constraints. Here the designers must consider what was learned about the problem during problem scoping. Design plans include clear communication of design ideas through media such as notebooks, blueprints, schematics, or storyboards. They also include details about the design, such as measurements, materials, colors, costs of materials, instructions for how things fit together, and sets of directions. Making the decision about which design idea to move forward involves considering the trade-offs of each design idea.

Once a clear design plan is in place, the designers must try the solution. Trying the solution includes developing a prototype (a testable model) based on the plan generated. The prototype might be something physical or a process to accomplish a goal. This component of design requires that the designers consider the risk involved in implementing the design. The prototype developed must be tested. Testing the solution includes conducting fair tests that verify whether the plan is a solution that is good enough to meet the client and end user needs and wants. Data need to be collected about the results of the tests of the prototype, and these data should be used to make evidence-based decisions regarding the design choices made in the plan. Here, the designers must again consider the criteria and constraints for the problem.

Using the data gathered from the testing, the designers must decide whether the solution is good enough to meet the client and end user needs and wants by assessment based on the criteria and constraints. Here, the designers must justify or reject design decisions based on the background research gathered while learning about the problem and on the evidence gathered during the testing of the solution. The designers must now decide whether to present the current solution to the client as a possibility or to do more iterations of design on the solution. If they decide that improvements need to be made to the solution, the designers must decide if there is more that needs to be understood about the problem, client, or end user; if another design idea should be tried; or if more planning needs to be conducted on the same design. One way or another, more work needs to be done.

Throughout the process of designing a solution to meet a client's needs and wants, designers work in teams and must communicate to each other, the client, and likely the end user. Teamwork is important in engineering design because multiple perspectives and differing skills and knowledge are valuable when working to solve problems. Communication is key to the success of the designed solution. Designers must communicate their ideas clearly using many different representations, such as text in an engineering notebook, diagrams, flowcharts, technical briefs, or memos to the client.

LEARNING CYCLE

The same format for the learning cycle is used in all grade levels throughout the STEM Road Map, so that students engage in a variety of activities to learn about phenomena in the modules thoroughly and have consistent experiences in the problem- and project-based learning modules. Expectations for learning by younger students are not as high as for older students, but the format of the progression of learning is the same. Students who have learned with curriculum from the STEM Road Map in early grades know

what to expect in later grades. The learning cycle consists of five parts—Introductory Activity/Engagement, Activity/Exploration, Explanation, Elaboration/Application of Knowledge, and Evaluation/Assessment—and is based on the empirically tested 5E model from BSCS (Bybee et al. 2006).

In the Introductory Activity/Engagement phase, teachers introduce the module challenge and use a unique approach designed to pique students' curiosity. This phase gets students to start thinking about what they already know about the topic and begin wondering about key ideas. The Introductory Activity/Engagement phase positions students to be confident about what they are about to learn, because they have prior knowledge, and clues them into what they don't yet know.

In the Activity/Exploration phase, the teacher sets up activities in which students experience a deeper look at the topics that were introduced earlier. Students engage in the activities and generate new questions or consider possibilities using preliminary investigations. Students work independently, in small groups, and in whole-group settings to conduct investigations, resulting in common experiences about the topic and skills involved in the real-world activities. Teachers can assess students' development of concepts and skills based on the common experiences during this phase.

During the Explanation phase, teachers direct students' attention to concepts they need to understand and skills they need to possess to accomplish the challenge. Students participate in activities to demonstrate their knowledge and skills to this point, and teachers can pinpoint gaps in student knowledge during this phase.

In the Elaboration/Application of Knowledge phase, teachers present students with activities that engage in higher-order thinking to create depth and breadth of student knowledge, while connecting ideas across topics within and across STEM. Students apply what they have learned thus far in the module to a new context or elaborate on what they have learned about the topic to a deeper level of detail.

In the last phase, Evaluation/Assessment, teachers give students summative feedback on their knowledge and skills as demonstrated through the challenge. This is not the only point of assessment (as discussed in the section on Embedded Formative Assessments), but it is an assessment of the culmination of the knowledge and skills for the module. Students demonstrate their cognitive growth at this point and reflect on how far they have come since the beginning of the module. The challenges are designed to be multidimensional in the ways students must collaborate and communicate their new knowledge.

STEM RESEARCH NOTEBOOK

One of the main components of the *STEM Road Map Curriculum Series* is the STEM Research Notebook, a place for students to capture their ideas, questions, observations, reflections, evidence of progress, and other items associated with their daily work. At the beginning of each module, the teacher walks students through the setup of the STEM

Research Notebook, which could be a three-ring binder, composition book, or spiral notebook. You may wish to have students create divided sections so that they can easily access work from various disciplines during the module. Electronic notebooks kept on student devices are also acceptable and encouraged. Students will develop their own table of contents and create chapters in the notebook for each module.

Each lesson in the *STEM Road Map Curriculum Series* includes one or more prompts that are designed for inclusion in the STEM Research Notebook and appear as questions or statements that the teacher assigns to students. These prompts require students to apply what they have learned across the lesson to solve the big problem or challenge for that module. Each lesson is designed to meaningfully refer students to the larger problem or challenge they have been assigned to solve with their teams. The STEM Research Notebook is designed to be a key formative assessment tool, as students' daily entries provide evidence of what they are learning. The notebook can be used as a mechanism for dialogue between the teacher and students, as well as for peer and self-evaluation.

The use of the STEM Research Notebook is designed to scaffold student notebooking skills across the grade bands in the *STEM Road Map Curriculum Series*. In the early grades, children learn how to organize their daily work in the notebook as a way to collect their products for future reference. In elementary school, students structure their notebooks to integrate background research along with their daily work and lesson prompts. In the upper grades (middle and high school), students expand their use of research and data gathering through team discussions to more closely mirror the work of STEM experts in the real world.

THE ROLE OF ASSESSMENT IN THE *STEM ROAD MAP CURRICULUM SERIES*

Starting in the middle years and continuing into secondary education, the word *assessment* typically brings grades to mind. These grades may take the form of a letter or a percentage, but they typically are used as a representation of a student's content mastery. If well thought out and implemented, however, classroom assessment can offer teachers, parents, and students valuable information about student learning and misconceptions that does not necessarily come in the form of a grade (Popham 2013).

The *STEM Road Map Curriculum Series* provides a set of assessments for each module. Teachers are encouraged to use assessment information for more than just assigning grades to students. Instead, assessments of activities requiring students to actively engage in their learning, such as student journaling in STEM Research Notebooks, collaborative presentations, and constructing graphic organizers, should be used to move student learning forward. Whereas other curriculum with assessments may include objective-type (multiple-choice or matching) tests, quizzes, or worksheets, we have intentionally avoided these forms of assessments to better align assessment strategies with teacher instruction and

student learning techniques. Since the focus of this book is on project- or problem-based STEM curriculum and instruction that focuses on higher-level thinking skills, appropriate and authentic performance assessments were developed to elicit the most reliable and valid indication of growth in student abilities (Brookhart and Nitko 2008).

Comprehensive Assessment System

Assessment throughout all STEM Road Map curriculum modules acts as a comprehensive system in which formative and summative assessments work together to provide teachers with high-quality information on student learning. Formative assessment occurs when the teacher finds out formally or informally what a student knows about a smaller, defined concept or skill and provides timely feedback to the student about his or her level of proficiency. Summative assessments occur when students have performed all activities in the module and are given a cumulative performance evaluation in which they demonstrate their growth in learning.

A comprehensive assessment system can be thought of as akin to a sporting event. Formative assessments are the practices: It is important to accomplish them consistently, they provide feedback to help students improve their learning, and making mistakes can be worthwhile if students are given an opportunity to learn from them. Summative assessments are the competitions: Students need to be prepared to perform at the best of their ability. Without multiple opportunities to practice skills along the way through formative assessments, students will not have the best chance of demonstrating growth in abilities through summative assessments (Black and Wiliam 1998).

Embedded Formative Assessments

Formative assessments in this module serve two main purposes: to provide feedback to students about their learning and to provide important information for the teacher to inform immediate instructional needs. Providing feedback to students is particularly important when conducting problem- or project-based learning because students take on much of the responsibility for learning, and teachers must facilitate student learning in an informed way. For example, if students are required to conduct research for the Activity/Exploration phase but are not familiar with what constitutes a reliable resource, they may develop misconceptions based on poor information. When a teacher monitors this learning through formative assessments and provides specific feedback related to the instructional goals, students are less likely to develop incomplete or incorrect conceptions in their independent investigations. By using formative assessment to detect problems in student learning and then acting on this information, teachers help move student learning forward through these teachable moments.

Formative assessments come in a variety of formats. They can be informal, such as asking students probing questions related to student knowledge or tasks or simply observing students engaged in an activity to gather information about student skills. Formative assessments can also be formal, such as a written quiz or a laboratory practical. Regardless of the type, three key steps must be completed when using formative assessments (Sondergeld, Bell, and Leusner 2010). First, the assessment is delivered to students so that teachers can collect data. Next, teachers analyze the data (student responses) to determine student strengths and areas that need additional support. Finally, teachers use the results from information collected to modify lessons and create learning environments that reinforce weak points in student learning. If student learning information is not used to modify instruction, the assessment cannot be considered formative in nature.

Formative assessments can be about content, science process skills, or even learning skills. When a formative assessment focuses on content, it assesses student knowledge about the disciplinary core ideas from the *Next Generation Science Standards* (*NGSS*) or content objectives from *Common Core State Standards for Mathematics* (*CCSS Mathematics*) or *Common Core State Standards for English Language Arts* (*CCSS ELA*). Content-focused formative assessments ask students questions about declarative knowledge regarding the concepts they have been learning. Process skills formative assessments examine the extent to which a student can perform science and engineering practices from the *NGSS* or process objectives from *CCSS Mathematics* or *CCSS ELA*, such as constructing an argument. Learning skills can also be assessed formatively by asking students to reflect on the ways they learn best during a module and identify ways they could have learned more.

Assessment Maps

Assessment maps or blueprints can be used to ensure alignment between classroom instruction and assessment. If what students are learning in the classroom is not the same as the content on which they are assessed, the resultant judgment made on student learning will be invalid (Brookhart and Nitko 2008). Therefore, the issue of instruction and assessment alignment is critical. The assessment map for this book (found in Chapter 3) indicates by lesson whether the assessment should be completed as a group or on an individual basis, identifies the assessment as formative or summative in nature, and aligns the assessment with its corresponding learning objectives.

Note that the module includes far more formative assessments than summative assessments. This is done intentionally to provide students with multiple opportunities to practice their learning of new skills before completing a summative assessment. Note also that formative assessments are used to collect information on only one or two learning objectives at a time so that potential relearning or instructional modifications can focus on smaller and more manageable chunks of information. Conversely, summative assessments in the module cover many more learning objectives, as they are traditionally used as final markers of student learning. This is not to say that information collected from summative assessments cannot or should not be used formatively. If teachers find that gaps in student

learning persist after a summative assessment is completed, it is important to revisit these existing misconceptions or areas of weakness before moving on (Black et al. 2003).

SELF-REGULATED LEARNING THEORY IN THE STEM ROAD MAP MODULES

Many learning theories are compatible with the STEM Road Map modules, such as constructivism, situated cognition, and meaningful learning. However, we feel that the self-regulated learning theory (SRL) aligns most appropriately (Zimmerman 2000). SRL requires students to understand that thinking needs to be motivated and managed (Ritchhart, Church, and Morrison 2011). The STEM Road Map modules are student centered and are designed to provide students with choices, concrete hands-on experiences, and opportunities to see and make connections, especially across subjects (Eliason and Jenkins 2012; NAEYC 2016). Additionally, SRL is compatible with the modules because it fosters a learning environment that supports students' motivation, enables students to become aware of their own learning strategies, and requires reflection on learning while experiencing the module (Peters and Kitsantas 2010).

The theory behind SRL (see Figure 2.2) explains the different processes that students engage in before, during, and after a learning task. Because SRL is a cyclical learning process, the accomplishment of one cycle develops strategies for the next learning cycle. This cyclic way of learning aligns with the various sections in the STEM Road Map lesson plans on Introductory Activity/ Engagement, Activity/Exploration, Explanation, Elaboration/Application of Knowledge, and Evaluation/Assessment. Since the students engaged in a module take on much of the responsibility for learning, this theory also provides guidance for teachers to keep students on the right track.

The remainder of this section explains how SRL theory is embedded within the five sections of each module and points out ways to

Figure 2.2. SRL Theory

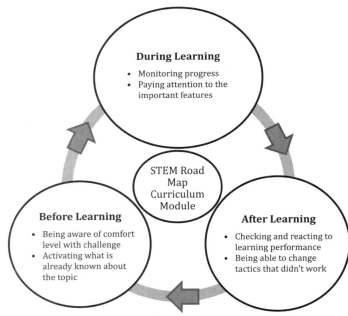

Source: Adapted from Zimmerman 2000.

support students in becoming independent learners of STEM while productively functioning in collaborative teams.

Before Learning: Setting the Stage

Before attempting a learning task such as the STEM Road Map modules, teachers should develop an understanding of their students' level of comfort with the process of accomplishing the learning and determine what they already know about the topic. When students are comfortable with attempting a learning task, they tend to take more risks in learning and as a result achieve deeper learning (Bandura 1986).

The STEM Road Map curriculum modules are designed to foster excitement from the very beginning. Each module has an Introductory Activity/Engagement section that introduces the overall topic from a unique and exciting perspective, engaging the students to learn more so that they can accomplish the challenge. The Introductory Activity also has a design component that helps teachers assess what students already know about the topic of the module. In addition to the deliberate designs in the lesson plans to support SRL, teachers can support a high level of student comfort with the learning challenge by finding out if students have ever accomplished the same kind of task and, if so, asking them to share what worked well for them.

During Learning: Staying the Course

Some students fear inquiry learning because they aren't sure what to do to be successful (Peters 2010). However, the STEM Road Map curriculum modules are embedded with tools to help students pay attention to knowledge and skills that are important for the learning task and to check student understanding along the way. One of the most important processes for learning is the ability for learners to monitor their own progress while performing a learning task (Peters 2012). The modules allow students to monitor their progress with tools such as the STEM Research Notebooks, in which they record what they know and can check whether they have acquired a complete set of knowledge and skills. The STEM Road Map modules support inquiry strategies that include previewing, questioning, predicting, clarifying, observing, discussing, and journaling (Morrison and Milner 2014). Through the use of technology throughout the modules, inquiry is supported by providing students access to resources and data while enabling them to process information, report the findings, collaborate, and develop 21st century skills.

It is important for teachers to encourage students to have an open mind about alternative solutions and procedures (Milner and Sondergeld 2015) when working through the STEM Road Map curriculum modules. Novice learners can have difficulty knowing what to pay attention to and tend to treat each possible avenue for information as equal (Benner 1984). Teachers are the mentors in a classroom and can point out ways for students to approach learning during the Activity/Exploration, Explanation, and

Elaboration/Application of Knowledge portions of the lesson plans to ensure that students pay attention to the important concepts and skills throughout the module. For example, if a student is to demonstrate conceptual awareness of motion when working on roller coaster research, but the student has misconceptions about motion, the teacher can step in and redirect student learning.

After Learning: Knowing What Works

The classroom is a busy place, and it may often seem that there is no time for self-reflection on learning. Although skipping this reflective process may save time in the short term, it reduces the ability to take into account things that worked well and things that didn't so that teaching the module may be improved next time. In the long run, SRL skills are critical for students to become independent learners who can adapt to new situations. By investing the time it takes to teach students SRL skills, teachers can save time later, because students will be able to apply methods and approaches for learning that they have found effective to new situations. In the Evaluation/Assessment portion of the STEM Road Map curriculum modules, as well as in the formative assessments throughout the modules, two processes in the after-learning phase are supported: evaluating one's own performance and accounting for ways to adapt tactics that didn't work well. Students have many opportunities to self-assess in formative assessments, both in groups and individually, using the rubrics provided in the modules.

The designs of the *NGSS* and *CCSS* allow for students to learn in diverse ways, and the STEM Road Map curriculum modules emphasize that students can use a variety of tactics to complete the learning process. For example, students can use STEM Research Notebooks to record what they have learned during the various research activities. Notebook entries might include putting objectives in students' own words, compiling their prior learning on the topic, documenting new learning, providing proof of what they learned, and reflecting on what they felt successful doing and what they felt they still needed to work on. Perhaps students didn't realize that they were supposed to connect what they already knew with what they learned. They could record this and would be prepared in the next learning task to begin connecting prior learning with new learning.

SAFETY IN STEM

Student safety is a primary consideration in all subjects but is an area of particular concern in science, where students may interact with unfamiliar tools and materials that may pose additional safety risks. It is important to implement safety practices within the context of STEM investigations, whether in a classroom laboratory or in the field. When you keep safety in mind as a teacher, you avoid many potential issues with the lesson while also protecting your students.

STEM safety practices encompass things considered in the typical science classroom. Ensure that students are familiar with basic safety considerations, such as wearing protective equipment (e.g., safety glasses or goggles and latex-free gloves) and taking care with sharp objects, and know emergency exit procedures. Teachers should learn beforehand the locations of the safety eyewash, fume hood, fire extinguishers, and emergency shut-off switch in the classroom and how to use them. Also be aware of any school or district safety policies that are in place and apply those that align with the work being conducted in the lesson. It is important to review all safety procedures annually.

STEM investigations should always be supervised. Each lesson in the modules includes teacher guidelines for applicable safety procedures that should be followed. Before each investigation, teachers should go over these safety procedures with the student teams. Some STEM focus areas such as engineering require that students can demonstrate how to properly use equipment in the maker space before the teacher allows them to proceed with the lesson.

Information about classroom science safety, including a safety checklist for science classrooms, general lab safety recommendations, and links to other science safety resources, is available at the Council of State Science Supervisors (CSSS) website at *www.csss-science. org/safety.shtml*. The National Science Teachers Association (NSTA) provides a list of science rules and regulations, including standard operating procedures for lab safety, and a safety acknowledgment form for students and parents or guardians to sign. You can access these resources at *http://static.nsta.org/pdfs/SafetyInTheScienceClassroom.pdf*. In addition, NSTA's Safety in the Science Classroom web page (*www.nsta.org/safety*) has numerous links to safety resources, including papers written by the NSTA Safety Advisory Board.

Disclaimer: The safety precautions for each activity are based on use of the recommended materials and instructions, legal safety standards, and better professional practices. Using alternative materials or procedures for these activities may jeopardize the level of safety and therefore is at the user's own risk.

REFERENCES

Bandura, A. 1986. *Social foundations of thought and action: A social cognitive theory.* Englewood Cliffs, NJ: Prentice-Hall.

Barell, J. 2006. *Problem-based learning: An inquiry approach.* Thousand Oaks, CA: Corwin Press.

Benner, P. 1984. *From novice to expert: Excellence and power in clinical nursing practice.* Menlo Park, CA: Addison-Wesley Publishing Company.

Black, P., C. Harrison, C. Lee, B. Marshall, and D. Wiliam. 2003. *Assessment for learning: Putting it into practice.* Berkshire, UK: Open University Press.

Black, P., and D. Wiliam. 1998. Inside the black box: Raising standards through classroom assessment. *Phi Delta Kappan* 80 (2): 139–148.

Blumenfeld, P., E. Soloway, R. Marx, J. Krajcik, M. Guzdial, and A. Palincsar. 1991. Motivating project-based learning: Sustaining the doing, supporting learning. *Educational Psychologist* 26 (3): 369–398.

Brookhart, S. M., and A. J. Nitko. 2008. *Assessment and grading in classrooms.* Upper Saddle River, NJ: Pearson.

Bybee, R., J. Taylor, A. Gardner, P. Van Scotter, J. Carlson, A. Westbrook, and N. Landes. 2006. *The BSCS 5E instructional model: Origins and effectiveness. http://science.education.nih.gov/houseofreps. nsf/b82d55fa138783c2852572c9004f5566/$FILE/Appendix?D.pdf.*

Eliason, C. F., and L. T. Jenkins. 2012. *A practical guide to early childhood curriculum.* 9th ed. New York: Merrill.

Johnson, C. 2003. Bioterrorism is real-world science: Inquiry-based simulation mirrors real life. *Science Scope* 27 (3): 19–23.

Krajcik, J., and P. Blumenfeld. 2006. Project-based learning. In *The Cambridge handbook of the learning sciences,* ed. R. Keith Sawyer, 317–334. New York: Cambridge University Press.

Lambros, A. 2004. *Problem-based learning in middle and high school classrooms: A teacher's guide to implementation.* Thousand Oaks, CA: Corwin Press.

Milner, A. R., and T. Sondergeld. 2015. Gifted urban middle school students: The inquiry continuum and the nature of science. *National Journal of Urban Education and Practice* 8 (3): 442–461.

Morrison, V., and A. R. Milner. 2014. Literacy in support of science: A closer look at cross-curricular instructional practice. *Michigan Reading Journal* 46 (2): 42–56.

National Association for the Education of Young Children (NAEYC). 2016. Developmentally appropriate practice position statements. *www.naeyc.org/positionstatements/dap.*

Peters, E. E. 2010. Shifting to a student-centered science classroom: An exploration of teacher and student changes in perceptions and practices. *Journal of Science Teacher Education* 21 (3): 329–349.

Peters, E. E. 2012. Developing content knowledge in students through explicit teaching of the nature of science: Influences of goal setting and self-monitoring. *Science and Education* 21 (6): 881–898.

Peters, E. E., and A. Kitsantas. 2010. The effect of nature of science metacognitive prompts on science students' content and nature of science knowledge, metacognition, and self-regulatory efficacy. *School Science and Mathematics* 110: 382–396.

Popham, W. J. 2013. *Classroom assessment: What teachers need to know.* 7th ed. Upper Saddle River, NJ: Pearson.

Ritchhart, R., M. Church, and K. Morrison. 2011. *Making thinking visible: How to promote engagement, understanding, and independence for all learners.* San Francisco, CA: Jossey-Bass.

Sondergeld, T. A., C. A. Bell, and D. M. Leusner. 2010. Understanding how teachers engage in formative assessment. *Teaching and Learning* 24 (2): 72–86.

Zimmerman, B. J. 2000. Attaining self-regulation: A social-cognitive perspective. In *Handbook of self-regulation,* ed. M. Boekaerts, P. Pintrich, and M. Zeidner, 13–39. San Diego: Academic Press.

PART 2

PATTERNS AND THE PLANT WORLD

STEM ROAD MAP MODULE

PATTERNS AND THE PLANT WORLD MODULE OVERVIEW

Andrea R. Milner, Vanessa B. Morrison, Janet B. Walton, Carla C. Johnson, and Erin Peters-Burton

THEME: The Represented World

LEAD DISCIPLINES: Mathematics and Science

MODULE SUMMARY

This module focuses on how changes in seasonal weather patterns relate to changes in the plant world. Students explore how Earth's movement around the Sun influences regional weather patterns on Earth and observe the changes in plant life that accompany the changing seasons. The class uses the steps of the engineering design process (EDP) to design and create a container garden in the culminating activity of the module, the Container Garden Design Challenge. Students add plants or seeds to the container they have designed so that they may observe plant growth cycles for an extended period of time. Students also collaboratively design an observation journal in which to record data about their garden, emphasizing observations, data collection, measurements, and graphic presentations of numerical data. Although the module lessons cover a span of five weeks, students will continue to reflect on their garden design, observe plant changes, and collect data after completing the module lessons (adapted from Koehler, Bloom, and Milner 2015).

ESTABLISHED GOALS AND OBJECTIVES

At the conclusion of this module, students will be able to do the following:

- Demonstrate conceptual awareness of how living things grow and change over the course of their lives:
  - Understand that plants change over time because of the changing seasons
  - Understand that certain plants grow in different regions (local and global)

- Understand that various habitats (local and global) are home to certain plants

- Use technology to gather research information and communicate

- Identify technological advances and tools that scientists use to learn about patterns and plants

- Describe and apply the EDP

- Explain concepts and analyze data they have observed about the life cycle of plants

- Understand local weather patterns and be able to make connections among weather, seasons, habitat, and the life cycle of plants

- Understand the influence gardening has on culture and society

CHALLENGE OR PROBLEM FOR STUDENTS TO SOLVE: CONTAINER GARDEN DESIGN CHALLENGE

As a culminating activity for the module, students participate in the Container Garden Design Challenge. In this challenge, students will design a window box or free-standing container garden, plant several different kinds of plants, and then observe their growth and follow their life cycle over an extended period of time (several months). Using a window box would allow students to easily observe the plants, but attaching a window box to your school structure may not be an option, so a free-standing container garden or raised-bed garden may be more appropriate. The kinds of plants chosen and the purpose of the class garden will vary depending on the geographic location, time of year, and space constraints. Students will make decisions about the plants during the planning process.

CONTENT STANDARDS ADDRESSED IN THIS STEM ROAD MAP MODULE

A full listing with descriptions of the standards this module addresses can be found in Appendix C. Listings of the particular standards addressed within lessons are provided in a table for each lesson in Chapter 4.

STEM RESEARCH NOTEBOOK

Each student should maintain a STEM Research Notebook, which will serve as a place for students to organize their work throughout this module (see p. 12 for more general discussion on setup and use of this notebook). All written work in the module should be included in the notebook, including records of students' thoughts and ideas, fictional accounts based on the concepts in the module, and records of student progress through the EDP. The notebooks may be maintained across subject areas, giving students the opportunity to see that although their classes may be separated during the school day, the knowledge they gain is connected. The lesson plans for this module contain STEM Research Notebook Entry sections (numbered 1–23) and templates for each notebook entry are included in Appendix A.

Emphasize to students the importance of organizing all information in a Research Notebook. Explain to them that scientists and other researchers maintain detailed Research Notebooks in their work. These notebooks, which are crucial to researchers' work because they contain critical information and track the researchers' progress, are often considered legal documents for scientists who are pursuing patents or wish to provide proof of their discovery process.

MODULE LAUNCH

Following agreed-upon rules for discussions, launch the module by holding a class discussion about plants and the seasons, asking students these questions:

- What are the four seasons?

- What causes the four seasons?

- How do the seasons affect plant development?

- How do plants change during the four seasons?

Then, have students explore the seasons by viewing a video about the seasons such as "The Four Seasons" by ABCmouse.com at *www.youtube.com/watch?v=K2tV69N0X8k*. Tell students that as part of their challenge in this module, they will design and build a class garden and observe the growth of the plants.

PREREQUISITE SKILLS FOR THE MODULE

Students enter this module with a wide range of preexisting skills, information, and knowledge. Table 3.1 (p. 26) provides an overview of prerequisite skills and knowledge that students are expected to apply in this module, along with examples of how they apply this knowledge throughout the module. Differentiation strategies are also provided for students who may need additional support in acquiring or applying this knowledge.

Table 3.1. Prerequisite Key Knowledge and Examples of Applications and Differentiation Strategies

Prerequisite Key Knowledge	Application of Knowledge by Students	Differentiation for Students Needing Additional Knowledge
Science • Understand cause and effect	*Science* • Determine how changes in the plant world and how changes on Earth (the seasons) and in the sky (daylight hours) affect plant development.	*Science* • Provide demonstrations of cause and effect (e.g., dropping egg [cause] and observing breakage [effect]), emphasizing that cause is why something happens, effect is what happens. • Read aloud picture books to class and have students identify cause and effect sequences. • Create a class T-chart to record causes and related effects students observe in the classroom and in literature.
Mathematics • Number sense	*Mathematics* • Record temperatures. • Measure dimensions for container garden. • Measure plant growth.	*Mathematics* • Model measurement techniques using standard and nonstandard units of measurement. • Read aloud nonfiction texts about temperature and measurement to class. • Provide opportunities for students to practice measurement in a variety of settings (e.g., in the classroom and outdoors).
Language and Inquiry Skills • Visualize • Make predictions • Ask and respond to questions	*Language and Inquiry Skills* • Make and confirm or reject predictions. • Share thought processes through notebooking, asking and responding to questions, and use of the engineering design process.	*Language and Inquiry Skills* • As a class, make predictions when reading fictional texts. • Model the process of using information and prior knowledge to make predictions. • Provide samples of notebook entries.

Continued

Table 3.1. (*continued*)

Prerequisite Key Knowledge	Application of Knowledge by Students	Differentiation for Students Needing Additional Knowledge
Speaking and Listening • Participate in group discussions	*Speaking and Listening* • Engage in collaborative group discussions in the development of the container garden and observation journal.	*Speaking and Listening* • Model speaking and listening skills. • Create a class list of good listening and good speaking practices. • Read picture books that feature collaboration and teamwork.

POTENTIAL STEM MISCONCEPTIONS

Students enter the classroom with a wide variety of prior knowledge and ideas, so it is important to be alert to misconceptions, or inappropriate understandings of foundational knowledge. These misconceptions can be classified as one of several types: "preconceived notions," opinions based on popular beliefs or understandings; "nonscientific beliefs," knowledge students have gained about science from sources outside the scientific community; "conceptual misunderstandings," incorrect conceptual models based on incomplete understanding of concepts; "vernacular misconceptions," misunderstandings of words based on their common use versus their scientific use; and "factual misconceptions," incorrect or imprecise knowledge learned in early life that remains unchallenged (NRC 1997, p. 28). Misconceptions must be addressed and dismantled for students to reconstruct their knowledge, and therefore teachers should be prepared to take the following steps:

- *Identify students' misconceptions.*

- *Provide a forum for students to confront their misconceptions.*

- *Help students reconstruct and internalize their knowledge, based on scientific models.*
  (NRC 1997, p. 29)

Keeley and Harrington (2010) recommend using diagnostic tools such as probes and formative assessment to identify and confront student misconceptions and begin the process of reconstructing student knowledge. Keeley's *Uncovering Student Ideas in Science* series contains probes targeted toward uncovering student misconceptions in a variety of areas and may be a useful resource for addressing student misconceptions in this module.

Some commonly held misconceptions specific to lesson content are provided with each lesson so that you can be alert for student misunderstanding of the science concepts presented and used during this module. The American Association for the Advancement of Science has also identified misconceptions that students frequently hold regarding various science concepts (see the links at *http://assessment.aaas.org/topics*).

SRL PROCESS COMPONENTS

Table 3.2 illustrates some activities in the Patterns and the Plant World module and how they align with the self-regulated learning (SRL) process before, during, and after learning.

Table 3.2. SRL Process Components

Learning Process Components	Example From Patterns and the Plant World Module	Lesson Number and Learning Component
BEFORE LEARNING		
Motivates students	Students document what they want to know about seasons in their STEM Research Notebooks during the group discussion and before viewing the video on the four seasons.	Lesson 1, Introductory Activity/ Engagement
Evokes prior learning	Students hold a class discussion on seasons and plant life cycles.	Lesson 1, Introductory Activity/ Engagement
DURING LEARNING		
Focuses on important features	Students learn about different types of plants through an interactive read-aloud.	Lesson 2, Introductory Activity/ Engagement
Helps students monitor their progress	Document student ideas about container gardening on a Know, Want to Know, Learned (KWL) chart. Students create STEM Research Notebook entries.	Lesson 2, Activity/ Exploration
AFTER LEARNING		
Evaluates learning	Students receive feedback on the rubric for observations on their listening and discussion skills, STEM Research Notebooks, and participation.	Lesson 3, Activity/ Exploration

Continued

Table 3.2. (*continued*)

Learning Process Components	Example From Patterns and the Plant World Module	Lesson Number and Learning Component
Takes account of what worked and what did not work	Students collaboratively decide what observations to make about the garden, using their experiences from observing plants on the school grounds as a guide. Students consider how to organize these observations to track how plants develop throughout the year and use this information to make decisions about the effectiveness of their garden design and possible improvements.	Lesson 3, Activity/ Exploration; Elaboration/ Application of Knowledge

STRATEGIES FOR DIFFERENTIATING INSTRUCTION WITHIN THIS MODULE

For the purposes of this curriculum module, differentiated instruction is conceptualized as a way to tailor instruction—including process, content, and product—to various student needs in your class. A number of differentiation strategies are integrated into lessons across the module. The problem- and project-based learning approach used in the lessons is designed to address students' multiple intelligences by providing a variety of entry points and methods to investigate the key concepts in the module (for example, investigating gardening from the perspectives of science and social issues via scientific inquiry, literature, journaling, and collaborative design). Differentiation strategies for students needing support in prerequisite knowledge can be found in Table 3.1 (p. 26). You are encouraged to use information gained about student prior knowledge during introductory activities and discussions to inform your instructional differentiation. Strategies incorporated into this lesson include flexible grouping, varied environmental learning contexts, assessments, compacting, tiered assignments and scaffolding, and mentoring. The following websites may be helpful resources for differentiated instruction:

- *http://steinhardt.nyu.edu/scmsAdmin/uploads/005/120/Culturally%20Responsive%20Differentiated%20Instruction.pdf*

- *http://educationnorthwest.org/sites/default/files/12.99.pdf*

Flexible Grouping. Students work collaboratively in a variety of activities throughout this module. Grouping strategies you might employ include student-led grouping, grouping students according to ability level or common interests, grouping students randomly, or grouping them so that students in each group have complementary strengths (for instance, one student might be strong in mathematics, another in art, and another in writing).

Varied Environmental Learning Contexts. Students have the opportunity to learn in various contexts throughout the module, including alone, in groups, in quiet reading and research-oriented activities, and in active learning through inquiry and design activities. In addition, students learn in a variety of ways, including through doing inquiry activities, journaling, reading a variety of texts, watching videos, participating in class discussion, and conducting web-based research.

Assessments. Students are assessed in a variety of ways throughout the module, including individual and collaborative formative and summative assessments. Students have the opportunity to produce work via written text, oral presentations, and modeling.

Compacting. Based on student prior knowledge, you may wish to adjust instructional activities for students who exhibit prior mastery of a learning objective. Since student work in science is largely collaborative throughout the module, this strategy may be most appropriate for mathematics, English language arts (ELA), or social studies activities.

Tiered Assignments and Scaffolding. Based on your awareness of student ability, understanding of concepts, and mastery of skills, you may wish to provide students with variations on activities by adding complexity to assignments or providing more or fewer learning supports for activities throughout the module. For instance, some students may need additional support in identifying key search words and phrases for web-based research or may benefit from cloze sentence handouts to enhance vocabulary understanding. Other students may benefit from expanded reading selections and additional reflective writing or from working with manipulatives and other visual representations of mathematical concepts. You may also work with your school librarian to compile a classroom database of research resources and supplementary readings for different reading levels and on a variety of topics related to the module challenge to provide opportunities for students to undertake independent reading. You may find the following website on scaffolding strategies helpful: *www.edutopia.org/blog/scaffolding-lessons-six-strategies-rebecca-alber.*

Mentoring. As group design teamwork becomes increasingly complex throughout the module, you may wish to have a resource teacher, older student, or volunteer work with groups that struggle to stay on task and collaborate effectively.

STRATEGIES FOR ENGLISH LANGUAGE LEARNERS

Students who are developing proficiency in English language skills require additional supports to simultaneously learn academic content and the specialized language associated with specific content areas. WIDA (2012) has created a framework for providing support to these students and makes available rubrics and guidance on differentiating instructional materials for English language learners (ELLs). In particular, ELL students may benefit from additional sensory supports such as images, physical modeling, and graphic representations of module content, as well as interactive support through collaborative work. This module incorporates a variety of sensory supports and offers ongoing opportunities for ELL students to work collaboratively.

When differentiating instruction for ELL students, you should carefully consider the needs of these students as you introduce and use academic language in various language domains (listening, speaking, reading, and writing) throughout this module. To adequately differentiate instruction for ELL students, you should have an understanding of the proficiency level of each student. The following five overarching preK–5 WIDA learning standards are relevant to this module:

- Standard 1: Social and Instructional language. Distinguish between information provided by pictures or other illustrations and information provided by the words in a text. Use the illustrations and details in a text to describe its key ideas.

- Standard 2: The language of Language Arts. Write opinion pieces in which they introduce the topic or name the book they are writing about, state an opinion, supply a reason for the opinion, and provide some sense of closure. Write informative/explanatory texts in which they name a topic, supply some facts about the topic, and provide some sense of closure. Write narratives in which they recount two or more appropriately sequenced events, include some details regarding what happened, use temporal words to signal event order, and provide some sense of closure.

- Standard 3: The language of Mathematics. Order three objects by length; compare the lengths of two objects indirectly by using a third object.

- Standard 4: The language of Science. An object's motion can be described by tracing and measuring its position over time.

- Standard 5: The language of Social Studies. Describe people, places, things, and events with relevant details, expressing ideas and feelings clearly.

SAFETY CONSIDERATIONS FOR THE ACTIVITIES IN THIS MODULE

Science activities in this module focus on growing plants from seeds and creating a container garden from recycled materials. Students should use caution when handling scissors, bottles, and cans. Sharp points or edges can cut or puncture skin, and bottles can break if not handled carefully. Also caution students not to eat seeds, as they may be treated with toxic chemicals. For more general safety guidelines, see the Safety in STEM section in Chapter 2 (p. 18).

DESIRED OUTCOMES AND MONITORING SUCCESS

The desired outcomes for this module are outlined in Table 3.3, along with suggested ways to gather evidence to monitor student success. For more specific details on desired outcomes, see the Established Goals and Objectives sections for the module and individual lessons.

Table 3.3. Desired Outcomes and Evidence of Success in Achieving Identified Outcomes

Desired Outcomes	Evidence of Success	
	Performance Tasks	Other Measures
Students understand and can demonstrate their knowledge about changes in the plant world and how changes on Earth (the seasons) and in the sky (daylight hours) affect plant development.	• Student teams develop and maintain a container garden. • Students each maintain a STEM Research Notebook with what they want to know, responses to questions, and observations. • Students design an observation journal to be used throughout the module and the rest of the year.	Students are assessed using the Observation, STEM Research Notebook, and Participation Rubric.

ASSESSMENT PLAN OVERVIEW AND MAP

Table 3.4 provides an overview of the major group and individual *products* and *deliverables*, or things that student teams will produce in this module, that constitute the assessment for this module. See Table 3.5 for a full assessment map of formative and summative assessments in this module.

Table 3.4. Major Products and Deliverables in Lead Disciplines for Groups and Individuals

Lesson	Major Group Products and Deliverables	Major Individual Products and Deliverables
1	• Terrific Seasonal Tree team presentations	• Terrific Seasonal Tree models • STEM Research Notebook entries #1–6 • Lesson Assessment questions
2	• Team research and presentations for Container Garden Design Challenge • Class decisions on garden design and final design drawing for garden	• STEM Research Notebook entries #7–15 • Student drawings for Container Garden Design Challenge • Lesson Assessment questions
3	• Completion of Container Garden Design Challenge • Creation of observation journal	• STEM Research Notebook entries #16–23 • Lesson Assessment questions
Ongoing	• Maintenance of garden • Redesign of and improvements to garden • Student presentations of garden design and output	• Data collected in observation journal

Table 3.5. Assessment Map for Patterns in the Plant World Module

Lesson	Assessment	Group/ Individual	Formative/ Summative	Lesson Objective Assessed
1	STEM Research Notebook *entries*	Individual	Formative	• Identify the four seasons. • Understand what causes the four seasons. • Describe how seasons affect plant development. • Name and identify the parts of a plant (roots, stem, leaves). • Provide examples of how plants adapt to the four seasons.
1	Participation in class weather observations and analysis *observation and participation rubric*	Individual	Formative	• Chart, graph, identify, describe, and analyze patterns of local weather to make connections among weather, seasons, habitat, and the life cycle of plants.

Continued

Table 3.5. (*continued*)

Lesson	Assessment	Group/ Individual	Formative/ Summative	Lesson Objective Assessed
1	Terrific Seasonal Tree Activity *performance task*	Group	Formative	• Understand that plants change over time because of the changing seasons. • Understand that certain plants grow in different regions (local and global). • Understand that various habitats (local and global) are home to certain plants.
1	Plant Drawing and Description *end of lesson assessment*	Individual	Summative	• Describe how seasons affect plant development. • Name and identify the parts of a plant (roots, stem, leaves). • Provide examples of how plants adapt to the four seasons.
2	STEM Research Notebook *entries*	Individual	Formative	• List the basic needs of all plants. • Describe the conditions necessary for growing plants in a container garden. • Explain what happens when all five basic needs are not met. • Identify where plants come from. • Recognize what types of plants grow from seeds. • Evaluate the influence plants have on culture and society.
2	Participation in class weather observations and analysis *observation and participation rubric*	Individual	Formative	• Chart, graph, identify, describe, and analyze patterns of local weather to make connections among weather, seasons, habitat, and the life cycle of plants.

Continued

Table 3.5. (*continued*)

Lesson	Assessment	Group/ Individual	Formative/ Summative	Lesson Objective Assessed
2	Team research *presentation*	Group	Formative	• Use technology tools to gather research information about the life cycle of plants. • Use technology to facilitate deeper conceptual understanding about the life cycle of plants. • Identify technological advances and tools that scientists use to learn about the life cycle of plants. • Design a container garden. • Explain concepts through the design of a journal to make observations of the life cycle of plants.
2	Plant Drawing and Description *end of lesson assessment*	Individual	Summative	• List the basic needs of all plants. • Describe the conditions necessary for growing plants in a container garden. • Explain what happens when all five basic needs are not met. • Identify where plants come from. • Recognize what types of plants grow from seeds.
3	STEM Research Notebook *entries*	Individual	Formative	• Determine what seeds need in order to sprout into seedlings. • Estimate how much time it will take for seeds to sprout into seedlings. • Estimate how much seedlings will grow each week.

Continued

Table 3.5. (*continued*)

Lesson	Assessment	Group/Individual	Formative/Summative	Lesson Objective Assessed
3	Participation in class weather observations and analysis *observation and participation rubric*	Individual	Formative	• Chart, graph, identify, describe, and analyze patterns of local weather to make connections among weather, seasons, habitat, and the life cycle of plants.
3	Observation Journal *performance task*	Group/Individual	Formative	• Explain concepts through the design of a journal to make observations of the life cycle of plants. • Observe, measure, quantify, and analyze data during the life cycle of plants.
3	Container Garden Design *performance task*	Group	Formative	• Construct a container garden.
3	Plant Drawing and Description *end of lesson assessment*	Individual	Summative	• Determine what seeds need in order to sprout into seedlings. • List and define the parts of a plant.

MODULE TIMELINE

Tables 3.6–3.10 (pp. 37–39) provide lesson timelines for each week of the module. These timelines are provided for general guidance only and are based on class times of approximately 30 minutes.

NATIONAL SCIENCE TEACHERS ASSOCIATION

Table 3.6. STEM Road Map Module Schedule for Week One

Day 1	Day 2	Day 3	Day 4	Day 5
Lesson 1 *Earth's Sensational Seasons* • Launch the module by introducing the module challenge with a discussion of seasons and plant life cycles. • Show video on the four seasons. • Begin class weather chart.	*Lesson 1* *Earth's Sensational Seasons* • Show video "One Year in 40 Seconds." • Dissect lima bean seed in the What's in a Seed? activity. • Plant lima bean and sunflower seeds in the How Does Your Garden Grow? activity.	*Lesson 1* *Earth's Sensational Seasons* • Conduct an interactive read-aloud of *Sunshine Makes the Seasons*, by Franklyn M. Branley. • Begin research for Terrific Seasonal Tree activity.	*Lesson 1* *Earth's Sensational Seasons* • Continue research for Terrific Seasonal Tree activity. • Create tree models.	*Lesson 1* *Earth's Sensational Seasons* • Give presentations for Terrific Seasonal Tree activity. • Begin class vocabulary chart.

Table 3.7. STEM Road Map Module Schedule for Week Two

Day 6	Day 7	Day 8	Day 9	Day 10
Lesson 1 *Earth's Sensational Seasons* • Conduct an interactive read-aloud of *Plants Live Everywhere*, by Mary Dodson Wade. • Review measurement skills in For Good Measure activity. • Discuss habitats and plant differences.	*Lesson 1* *Earth's Sensational Seasons* • Discuss the type of gardening students will be doing. • Discuss different types of plants. • Answer the Lesson Assessment questions.	*Lesson 2* *Our Container Garden: Design Time* • Discuss basic needs of living things. • Conduct an interactive read-aloud of *Seeds*, by Vijaya Khisty Bodach.	*Lesson 2* *Our Container Garden: Design Time* • Conduct an interactive read-aloud of *Trees, Weeds, and Vegetables: So Many Kinds of Plants!* by Mary Dodson Wade.	*Lesson 2* *Our Container Garden: Design Time* • Introduce engineering design process (EDP). • Show video "Container Gardening with Kids."

Table 3.8. STEM Road Map Module Schedule for Week Three

Day 11	Day 12	Day 13	Day 14	Day 15
Lesson 2 *Our Container Garden: Design Time* • Brainstorm ideas about assigned topics on container gardens and begin team research on container gardens.	*Lesson 2* *Our Container Garden: Design Time* • Conduct team research on container gardens. • Begin to share team research with the class.	*Lesson 2* *Our Container Garden: Design Time* • Share team research with the class. • Make decisions about the type of garden the class will build. • Create a class list of necessary materials.	*Lesson 2* *Our Container Garden: Design Time* • Create the class container garden design. • Conduct interactive read-aloud of *Plant Secrets*, by Emily Goodman. • Discuss the impact gardening has on culture.	*Lesson 2* *Our Container Garden: Design Time* • Analyze and organize the weather data students have been recording. • Make a supply list for the container garden. • Optional: Design a watering system.

Table 3.9. STEM Road Map Module Schedule for Week Four

Day 16	Day 17	Day 18	Day 19	Day 20
Lesson 2 *Our Container Garden: Design Time* • Make a plan for providing necessary supplies. • Conduct an interactive read-aloud of *Flower Garden*, by Eve Bunting.	*Lesson 2* *Our Container Garden: Design Time* • Show video "Grow—Episode 6: Thrive" from Whole Foods Market. • Continue working on the list of supplies.	*Lesson 2* *Our Container Garden: Design Time* • Come up with ideas about what to do with the garden's harvest. • Discuss philanthropy and cultural and societal implications of gardening.	*Lesson 2* *Our Container Garden: Design Time* • Discuss careers associated with gardening and weather. • Answer the Lesson Assessment questions.	*Lesson 3* *Our Container Garden: Planting Time* • Review progress through the Define, Learn, and Plan phases of the EDP. • Conduct an interactive read-aloud of *From Seed to Plant*, by Gail Gibbons.

Table 3.10. STEM Road Map Module Schedule for Week Five

Day 21	Day 22	Day 23	Day 24	Day 25
Lesson 3 *Our Container Garden:* *Planting Time* • Make predictions about plant growth.	*Lesson 3* *Our Container Garden:* *Planting Time* • Start to build the container garden.	*Lesson 3* *Our Container Garden:* *Planting Time* • Finish building the container garden. • Observe plants on school grounds.	*Lesson 3* *Our Container Garden:* *Planting Time* • Create observation journal for garden. • Conduct an interactive read-aloud of *How a Seed Grows*, by Helene J. Jordan.	*Lesson 3* *Our Container Garden:* *Planting Time* • Create a maintenance plan for the garden. • Answer the Lesson Assessment questions. **Ongoing** • Observe and record data about the plants and maintain the garden. • Compare plant growth with predictions, connect plant health and ease of maintenance with garden design, and make design modifications. • Share garden design and plant data with other classes, parents, or community members.

RESOURCES

The media specialist can help you locate resources for students to view and read about plants, weather, and related content. Special educators and reading specialists can help find supplemental sources for students needing extra support in reading and writing. Additional resources may be found online. Community resources for this module may include botanists and horticulturists.

REFERENCES

Keeley, P., and R. Harrington. 2010. *Uncovering student ideas in physical science, volume 1: 45 new force and motion assessment probes.* Arlington, VA: NSTA Press.

Koehler, C., M. A. Bloom, and A. R. Milner. 2015. The STEM Road Map for grades K–2. In *STEM Road Map: A framework for integrated STEM education,* ed. C. C. Johnson, E. E. Peters-Burton, and T. J. Moore, 41–67. New York: Routledge. *www.routledge.com/products/9781138804234.*

National Research Council (NRC). 1997. *Science teaching reconsidered: A handbook.* Washington, DC: National Academies Press.

WIDA. 2012. 2012 amplification of the English language development standards: Kindergarten–grade 12. *https://wida.wisc.edu/teach/standards/eld.*

PATTERNS AND THE PLANT WORLD LESSON PLANS

Andrea R. Milner, Vanessa B. Morrison, Janet B. Walton, Carla C. Johnson, and Erin Peters-Burton

Lesson Plan 1: Earth's Sensational Seasons

In this lesson, students explore the concept of the four seasons and how the changing seasons affect plant development. They will begin to understand that seasonal changes are caused by Earth's tilt as it rotates around the Sun. Students focus on seasonal changes in trees and investigate the changes in one kind of tree that grows in their home region.

ESSENTIAL QUESTIONS

- What are the four seasons?
- What causes the four seasons?
- How do the seasons affect plant development?
- How do plants change during the four seasons?

ESTABLISHED GOALS AND OBJECTIVES

At the conclusion of this lesson, students will be able to do the following:

- Identify the four seasons
- Understand what causes the four seasons
- Describe how seasons affect plant development
- Name and identify the parts of a plant (roots, stem, leaves)
- Provide examples of how plants adapt to the four seasons
- Demonstrate conceptual awareness of how living things grow and change over the course of their lives:
  - Understand that plants change over time due to the changing seasons

- Understand that certain plants grow in different regions (local and global)

- Understand that various habitats (local and global) are home to certain plants

- Use technology tools to gather research information about the life cycle of plants

- Chart, graph, identify, describe, and analyze patterns of local weather to make connections among weather, seasons, habitat, and the life cycle of plants

- Discuss the influence gardening has on culture and society

TIME REQUIRED

- 7 days (approximately 30 minutes each day; see Tables 3.6–3.7, p. 37)

MATERIALS

Required Materials for Lesson 1

- STEM Research Notebooks (1 per student, see p. 24 for STEM Research Notebook information)

- Computer with internet access for viewing videos

- Smartphones or tablets for student video recording

- Weather chart for the entire class (create or purchase) or handouts for each student (attached at the end of this lesson)

- Images of different types of landscapes and plants available (e.g., a golf course, a vegetable garden, a forest)

- Images of various types of habitats (e.g., forest, desert, grassland, tundra, fresh water stream, ocean)

- Several different types of plant seedlings, including at least one each of a vegetable, fruit, flower from bulb, tree, and grass

- Pots of various sizes to accommodate the plants' growth for several months

- Books
  - *Sunshine Makes the Seasons*, by Franklyn M. Branley (HarperCollins, 2016)
  - *Plants Live Everywhere,* by Mary Dodson Wade (Enslow Elementary, 2009)

- Chart paper

- Markers

- Plain white paper (1 sheet per student)

- Lined writing paper (1 sheet per student)

- Pencils (1 per student)

- Crayons for use in STEM Research Notebook entries (1 set per student)

- Map or globe

- Safety glasses or goggles, nonlatex aprons, and nonlatex gloves

Additional Materials for What's in a Seed? (per student)

- 1 lima bean seed

- 1 paper towel

- 1 toothpick

Additional Materials for How Does Your Garden Grow? (per team of 2 students)

- Paper towels sufficient to cover desk area

- 1 lima bean seed

- 2 or 3 sunflower seeds

- 2 plastic cups with holes in bottom

- 2 plastic lids (must fit under plastic cups)

- Potting soil

- Plastic cup of water (either a cup with 6 ounces of water or an empty cup and access to water)

Additional Materials for Terrific Seasonal Tree (per student)

- 1 paper bag

- 2 plastic grocery bags (or 2–3 pieces of paper)

- Various colors of construction paper (green, brown, red, yellow, orange)

- Glue

- Scissors

Additional Materials for For Good Measure

- Pencil
- Piece of string
- Rulers (1 per student)

SAFETY NOTES

1. Remind students that personal protective equipment (safety glasses or goggles, aprons, and gloves) must be worn during all phases of this inquiry activity.

2. Caution students not to eat seeds, which may have been treated with toxic chemicals such as herbicides, pesticides, or fungicides.

3. Students should use caution when handling toothpicks and scissors, as the sharp points and blades can cut or puncture skin.

4. Immediately wipe up any spilled water or soil on the floor to avoid a slip-and-fall hazard.

5. Have students wash hands with soap and water after the activity is completed.

CONTENT STANDARDS AND KEY VOCABULARY

Table 4.1 lists the content standards from the *Next Generation Science Standards (NGSS), Common Core State Standards (CCSS)*, National Association for the Education of Young Children (NAEYC), and the Framework for 21st Century Learning that this lesson addresses, and Table 4.2 (p. 48) presents the key vocabulary. Vocabulary terms are provided for both teacher and student use. You may choose to introduce some or all of the terms to students.

Table 4.1. Content Standards Addressed in STEM Road Map Module Lesson 1

NEXT GENERATION SCIENCE STANDARDS

PERFORMANCE EXPECTATIONS

- 1-ESS1-1. Use observations of the sun, moon, and stars to describe patterns that can be predicted.

- 1-ESS1-2. Make observations at different times of year to relate the amount of daylight to the time of year.

SCIENCE AND ENGINEERING PRACTICES

Planning and Carrying Out Investigations

Planning and carrying out investigations to answer questions or test solutions to problems in K–2 builds on prior experiences and progresses to simple investigations, based on fair tests, which provide data to support explanations or design solutions.

- Plan and conduct investigations collaboratively to produce evidence to answer a question.

Analyzing and Interpreting Data

Analyzing data in K–2 builds on prior experiences and progresses to collecting, recording, and sharing observations.

- Use observations (firsthand or from media) to describe patterns in the natural world in order to answer scientific questions.

DISCIPLINARY CORE IDEAS

ESS1.A: The Universe and Its Stars

- Patterns of the motion of the sun, moon, and stars in the sky can be observed, described, and predicted.

ESS1.B: Earth and the Solar System

- Seasonal patterns of sunrise and sunset can be observed, described, and predicted.

CROSSCUTTING CONCEPT

Patterns

- Patterns in the natural world can be observed, used to describe phenomena, and used as evidence.

COMMON CORE STATE STANDARDS FOR MATHEMATICS

MATHEMATICAL PRACTICES

- MP1. Make sense of problems and persevere in solving them.

- MP2. Reason abstractly and quantitatively.

- MP3. Construct viable arguments and critique the reasoning of others.

Continued

Table 4.1. (*continued*)

- MP4. Model with mathematics.
- MP5. Use appropriate tools strategically.
- MP6. Attend to precision.
- MP7. Look for and make use of structure.
- MP8. Look for and express regularity in repeated reasoning.

MATHEMATICAL CONTENT

- 1.NBT.B.3. Compare two two-digit numbers based on meanings of the tens and ones digits, recording the results of comparisons with the symbols >, =, and <.
- 1.MD.A.1. Order three objects by length; compare the lengths of two objects indirectly by using a third object.
- 1.MD.C.4. Organize, represent, and interpret data with up to three categories; ask and answer questions about the total number of data points, how many in each category, and how many more or less are in one category than in another.
- 1.OA.A.1. Use addition and subtraction within 20 to solve word problems involving situations of adding to, taking from, putting together, taking apart, and comparing, with unknowns in all positions.
- 1.OA.A.2. Solve word problems that call for addition of three whole numbers whose sum is less than or equal to 20, e.g., by using objects, drawings, and equations with a symbol for the unknown number to represent the problem.

COMMON CORE STATE STANDARDS FOR ENGLISH LANGUAGE ARTS

READING STANDARDS

- RI.1.1. Ask and answer questions about key details in a text.
- RI.1.2. Identify the main topic and retell key details of a text.
- RI.1.3. Describe the connection between two individuals, events, ideas, or pieces of information in a text.
- RI.1.7. Use the illustrations and details in a text to describe its key ideas.

WRITING STANDARDS

- W.1.2. Write informative/explanatory texts in which they name a topic, supply some facts about the topic, and provide some sense of closure.
- W.1.7. Participate in shared research and writing.
- W.1.8. With guidance and support from adults, recall information from experiences or gather information from provided sources to answer a question.

Continued

Table 4.1. (*continued*)

SPEAKING AND LISTENING STANDARDS

- SL.1.1. Participate in collaborative conversations with diverse partners about *grade 1 topics and texts* with peers and adults in small and larger groups.

- SL.1.1.A. Follow agreed-upon rules for discussions.

- SL.1.1.B. Build on others' talk in conversations by responding to the comments of others through multiple exchanges.

- SL.1.1.C. Ask questions to clear up any confusion about the topics and texts under discussion.

- SL.1.3. Ask and answer questions about what a speaker says in order to gather additional information or clarify something that is not understood.

- SL.1.5. Add drawings or other visual displays to descriptions when appropriate to clarify ideas, thoughts, and feelings.

NATIONAL ASSOCIATION FOR THE EDUCATION OF YOUNG CHILDREN STANDARDS

- 2.E.1. Arrange firsthand, meaningful experiences that are intellectually and creatively stimulating, invite exploration and investigation, and engage children's active, sustained involvement by providing a rich variety of material, challenges, and ideas.

- 2.F.3. Extend the range of children's interests and the scope of their thought, present novel experiences and introduce stimulating ideas, problems, experiences, or hypotheses.

- 2.F.6. Enhance children's conceptual understanding through various strategies, including intensive interview and conversation, encourage children to reflect on and "revisit" their experiences.

- 2.G.2. Scaffolding takes on a variety of forms.

- 2.J.1. Incorporate a wide variety of experiences, materials and equipment, and teaching strategies to accommodate the range of children's individual differences in development, skills and abilities, prior experiences, needs, and interests.

- 3.A.1. Teachers consider what children should know, understand, and be able to do across the domains.

FRAMEWORK FOR 21ST CENTURY LEARNING

- Interdisciplinary Themes; Learning and Innovation Skills; Information, Media, and Technology Skills; Life and Career Skills.

Table 4.2. Key Vocabulary for Lesson 1

Key Vocabulary	Definition
climate	the weather conditions in an area over an extended period of time
environment	the conditions and objects including living things that are in our surroundings
flower	the part of a plant that produces seeds
fruit	the part of a plant that contains seeds and can be eaten as food
germinate	to start to grow
habitat	a place in nature where plants, animals, and people grow and live
leaf	the flat part of a plant that is attached to the stem (plural: leaves)
plant	a living thing that has roots, stems, leaves, flowers, and often fruits
produce	food products that have been grown, such as fruits and vegetables
root	the part of a plant that grows under the soil and takes water to the other parts of the plant
seasons	the four main periods of the year, which have different weather patterns and hours of daylight
seed	the part of a plant that can develop into another plant
seedling	a young plant that grows from a seed
soil	the dirt in which plants grow
stem	the main body of a plant
sunlight	light that comes from the Sun
vegetable	a part of a plant that is used as food
water	a liquid that has no color, taste, or smell
weather	the daily conditions over a particular area, including temperature, precipitation, cloud cover, and air pressure

TEACHER BACKGROUND INFORMATION

First graders are able to make connections across multiple content areas (STEM and English language arts [ELA]), as well as the various developmental domains (physical, social and emotional, personality, cognitive, and language). Incorporating students' prior knowledge with developmentally appropriate instruction will enable them to make these connections. Throughout this module, you should support and facilitate the advancement of these content areas and developmental domains within each student. For information about how formative assessments can be used to connect student prior experiences with classroom instruction, see the STEM Teaching Tools resource "Making Science Instruction Compelling for All Students: Using Cultural Formative Assessment to Build on Learner Interest and Experience" at *http://stemteachingtools.org/pd/sessionc*.

Seasonal Changes and Weather Patterns

This module focuses on seasonal changes and plant life cycles. For information about seasons and the changing distribution of the Sun's energy across Earth, see the National Oceanic and Atmospheric Administration's Changing Seasons web page at *www. education.noaa.gov/Climate/Changing_Seasons.html*. For information about plant life cycles, see National Geographic Kids' The Life Cycle of Flowering Plants web page at *www. natgeokids.com/za/discover/science/nature/the-life-cycle-of-flowering-plants*.

In this module, the class will track daily weather conditions and look for patterns over time. Since students may have heard the term *climate*, be prepared to distinguish between *weather* as daily atmospheric conditions in an area and *climate* as the weather conditions in an area over an extended period of time. You should emphasize to students that the class's observations do not span enough time to reflect climate patterns or changes, since interpreting these patterns require collecting and analyzing data for many years. For more information on the distinction between weather and climate, see NASA's What's the Difference Between Weather and Climate? web page at *www.nasa. gov/mission_pages/noaa-n/climate/climate_weather.html*.

This lesson focuses on changes in trees through the seasons. The Minnesota Department of Natural Resources provides an interactive poster called "Trees for All Seasons," which depicts trees in various seasons, at *www.dnr.state.mn.us/forestry/education/treeforallseasons/index.html*.

Career Connections

You may wish to introduce careers associated with weather and plants during this module, such as the following (adapted from Koehler, Bloom, and Milner 2015):

- agriculturist

- astronomer

- botanist

- ecologist

- geographer

- horticulturist

- mathematician

- meteorologist

- scientist

For more information about these and other careers, see the Bureau of Labor Statistics' *Occupational Outlook Handbook* at *www.bls.gov/ooh/home.htm*.

In this module, students are introduced to the idea that engineers and those in other STEM careers work together in teams to solve problems. Students will experience working in teams and in pairs as they progress through a simple scientific process, including predicting, observing, and explaining phenomena related to plant growth and seasonal changes in this lesson. This introduction to teamwork sets the stage for students' use of the engineering design process (EDP) later in the module.

Know, Want to Know, Learned Charts

Throughout this module, you will track student knowledge on Know, Want to Know, Learned (KWL) charts. These charts will be used to access and assess student prior knowledge, encourage students to think critically about the topic under discussion, and track student learning throughout the module. Each chart should consist of three columns, labeled "What We Know," "What We Want to Know," and "What We Learned." Write the topic at the top of each chart. It may be helpful to post these charts in a prominent place in the classroom so that students can refer to them throughout the module. Students will include their personal know, want to know, and learned reflections in their STEM Research Notebook entries.

Interactive Read-Alouds

This module also uses interactive read-alouds to engage students, access their prior knowledge, develop student background knowledge, and introduce topical vocabulary. These read-alouds expose children to teacher-read literature that may be beyond their independent reading levels but is consistent with their listening level. Interactive read-alouds may incorporate a variety of techniques, and you can find helpful information regarding these techniques at the following websites:

- *www.readingrockets.org/article/repeated-interactive-read-alouds-preschool-and-kindergarten*

- *www.k5chalkbox.com/interactive-read-aloud.html*

- *www.readwritethink.org/professional-development/strategy-guides/teacher-read-aloud-that-30799.html*

In general, interactive read-alouds provide opportunities for students to share prior knowledge and experiences, interact with the text and concepts introduced therein, launch conversations about the topics introduced, construct meaning, make predictions, and draw comparisons. You may wish to mark places within the texts to pause to ask for student experiences, predictions, or other ideas. Each reading experience should focus on an ongoing interaction between students and the text, including the following:

- Allow students to share personal stories throughout the reading.

- Ask students to predict throughout the story.

- Allow students to add new ideas from the book to the KWL chart and their STEM Research Notebooks.

- Allow students to add new words from the book to the vocabulary chart and their STEM Research Notebooks.

The materials list for each lesson includes the books for interactive read-alouds that you will use in that lesson. A list of suggested books for additional reading can be found at the end of this chapter (see p. 96).

COMMON MISCONCEPTIONS

Students will have various types of prior knowledge about the concepts introduced in this lesson. Table 4.3 outlines some common misconceptions students may have concerning these concepts. Because of the breadth of students' experiences, it is not possible to anticipate every misconception that students may bring as they approach this lesson. Incorrect or inaccurate prior understanding of concepts can influence student learning in the future, however, so it is important to be alert to misconceptions such as those presented in the table.

Table 4.3. Common Misconceptions About the Concepts in Lesson 1

Topic	Student Misconception	Explanation
Seasons	The seasons are caused by Earth's varying distance from the Sun (for example, Earth is closer to the Sun in the summer and farther away in the winter).	The orbit of Earth around the Sun is circular; seasons are caused by the tilt of Earth's axis.
	Seasons happen at the same time everywhere on Earth.	The tilt of Earth's axis means that when the Northern Hemisphere is tilted toward the Sun, the daylight hours are longer, resulting in the summer season.

PREPARATION FOR LESSON 1

Review the Teacher Background Information, assemble the materials for the lesson, duplicate the student handouts, and preview the videos recommended in the Learning Components section that follows. Present students with their STEM Research Notebooks and explain how these will be used (see p. 24). Templates for the STEM Research Notebook are provided in Appendix A, and a rubric for observations, student participation, and STEM Research Notebook entries is provided in Appendix B.

STEM Research Notebook Entry #5 provides a template for students to record vocabulary words. You may wish to use this template throughout the module for students to record definitions and illustrations of key vocabulary words. The template provides space for definitions and illustrations of three words. If you introduce more than three vocabulary words in a lesson you should make multiple copies of the template for each student.

Students will track the local weather throughout the module. Create or purchase a weather chart for class use that will accommodate your needs based on your local weather patterns (see p. 54 for a list of what students should include on the chart). A sample of weather symbols is provided at the end of this lesson. You may also wish to have students track the weather conditions individually; a sample weather chart is provided at the end of this lesson. You should adjust the weather-tracking methods to your region and the time of year. For example, if you live in an area where the weather is consistently warm, you may wish to focus on daily weather patterns, such as the change in temperature over the day and changes in the amount of cloud cover each day. Alternatively, if you are using this module in a place and during a time period in which seasonal weather changes are occurring, you may wish to incorporate an analysis of weekly trends, such as the number of days with a high temperature over 70 degrees each week.

In this lesson, students will plant lima bean and sunflower seeds. This activity requires plastic cups with holes for drainage in the bottom. You should create these holes in the cups in advance. Since lima bean seeds have a thick seed coat, they will take some time to germinate, so you should soak these seeds overnight to increase the speed of germination. Each pair of students will plant one lima bean seed and dissect a second. You may wish to put appropriate amounts of potting soil (enough to fill the plastic cups three-quarters full) for each team in sealable plastic bags or containers. Students will observe and measure their plants' growth over the course of the module. Create a class chart for each team of students to record plant growth on a weekly basis.

The class will choose one kind of tree that grows in your region to investigate in the Terrific Seasonal Tree activity. The Arbor Day Foundation provides a database of suggested trees by region at *https://shop.arborday.org/content.aspx?page=tree-nursery*. You may wish to provide students with a limited number of options to choose from that are appropriate for your area. Student teams will research the effect that each season has on their chosen tree species. Conducting targeted internet searches regarding this information may be difficult for first graders, so you may wish to print information and pictures in advance for student research or have reference books available in the classroom. You also may wish to gather a variety of images of different types of landscapes and plants available (e.g., a golf course, a vegetable garden, a forest) to prompt students to name different types of plants. Books, gardening magazines, or an internet search should yield a variety of images.

You should acquire several types of plant seedlings (see the materials list on p. 42) for class observation throughout the module. You should also have on hand pots of various sizes to accommodate these plants' growth over several months and enough potting soil to plant these. You may wish to plant them in front of the class, so that students can view the roots or bulbs of each plant. The class will observe these plants over time, noting changes such as growth in height, number of leaves, and development of buds. Create a class chart to record observations for each plant over the course of several months.

LEARNING COMPONENTS
Introductory Activity/Engagement

Connection to the Challenge: Begin each day of this lesson by directing students' attention to the module challenge, the Container Garden Design Challenge:

> *Your town's leaders have noticed that many of the fruits, vegetables, and flowers sold in your grocery stores were grown far away from your local community. They think that growing more of these items locally will provide produce that is fresher and less expensive. Because of this, the town's leaders are looking for ways for families, schools, and businesses to use space creatively to grow fruits, vegetables, and flowers for members of the community to use. Your class has been challenged to create a garden to grow produce or other plants at your school as a model for the community of how this can be done and how your local weather patterns affect plants.*

Tell students that they will design and build a class garden and observe the growth of the plants. (The plants and purpose of your garden will vary according to your geographic location, time of year, and space constraints; students will make decisions about plants during the planning process.) Tell students that to understand what plants will grow well in your area and at this time of year, they will need to do some research about how plants grow, the weather in your area, and how the seasons affect plants. Hold a brief class discussion of how students' learning in the previous days' lessons contributed to their ability to complete the challenge. You may wish to create a class list of key ideas on chart paper.

Mathematics and Science Classes: Introduce seasons and plant life cycles with a class discussion. Following agreed-upon rules for discussions, ask students:

- What are the four seasons?

- What causes the four seasons?

- How do the seasons affect plant development?

- How do plants change during the four seasons?

As students share their ideas, chart student responses on a KWL chart in the Know column. Then, ask students what questions they have about the seasons, recording these questions in the Want to Know column.

Show a video about the seasons such as "The Four Seasons" by ABCmouse.com at *www.youtube.com/watch?v=K2tV69N0X8k*. The video shown should include images of conditions for each of the four seasons as well as descriptions of weather conditions common to the seasons.

STEM Research Notebook Entry #1

After viewing the video, ask students to reflect in their STEM Research Notebooks on what they learned about seasons from the discussion and video, using both words and pictures.

Introduce students to the weather chart that the class will maintain throughout the module. Ask students for their ideas about why understanding weather might be useful for gardening (e.g., knowing what the weather conditions are like will help them choose what plants to grow). Introduce the term *environment* to students and emphasize to them that weather is an important part of the environment in which they live and in which plants live. Beginning on Day 1, at the start of every class, students will observe and chart, graph, identify, describe, and analyze patterns of local weather to make connections among daily weather, seasons, and plant life cycles. These observations will be made throughout the module to enable students to relate different times of year to the amount of daylight. (Emphasis is on relative comparisons of the amount of daylight in the winter with the amount in the spring or fall, not quantifying the hours or time of daylight.) Working as a class, students should observe and chart the following daily:

- Whether each day is warmer than, colder than, or the same temperature as the previous day

- Descriptions of the weather (such as sunny, cloudy, rainy, and warm)

- Numbers of sunny, windy, and rainy days in a month

- Observations of the Sun, Moon, and stars to describe patterns that can be predicted (e.g., pattern analysis and predictions could include that it is usually cooler in the morning than in the afternoon; different months have different numbers of sunny days and cloudy days; the Sun and Moon appear to rise in one part of the sky, move across the sky, and set; stars other than our Sun are visible at night but not during the day)

Have students continue their exploration of seasons by viewing a timelapse video such as "One Year in 40 Seconds" at *www.youtube.com/watch?v=lmIFXIXQQ_E*. After students respond to STEM Research Notebook Entry #2, document student ideas on the KWL chart.

STEM Research Notebook Entry #2

After viewing the video, ask students to reflect in their STEM Research Notebooks on what they learned about the progression of seasons, using both words and pictures.

Discuss the time scale for plant growth with students, explaining that it may take many weeks or months for plants to reach their full size. Tell students that they will be growing plants as part of the module challenge. They will plant lima bean and sunflower seeds in this lesson and observe the plants' growth over the course of the module. They will also design and build a container garden. They may transplant their plants that they grew from seed into their garden later in the module.

What's in a Seed?

First, however, students will dissect a lima bean seed to learn about the parts of a seed. Give each student one soaked lima bean seed, a paper towel, and a toothpick. Have students observe the bean seed and make predictions about what is inside the seed, recording student predictions on chart paper. Next, have students rub the seed gently between their fingers so that the outer layer comes off. (They may use toothpicks to aid in the removal if necessary.) Tell students that this outer layer is called the seed coat. Direct students' attention to the slit that runs down the middle of the seed and have them gently split the seed open at the slit, using the toothpick if necessary. Ask students what they see inside the seed. Tell students that the small green piece they see is the baby plant, called an embryo. The other material inside the seed is the food supply. Have each student draw their bean and label the parts: seed coat, embryo, and food supply.

How Does Your Garden Grow?

Next, have students work with partners to plant seeds. Give each pair of students the items on the materials list for this activity.

Have students fill their plastic cups about one-half to two-thirds full with potting soil, then add one lima bean seed to one of the cups and two to three sunflower seeds to the other cup and cover the seeds with about 1 inch of soil. Students should use markers to label the cups with the name of the type of seeds in each. Students should place their planted seeds in a sunny spot, with a plastic lid underneath each cup, and then moisten the soil with a small amount of water. Introduce the term *germinate* to students and hold a class discussion about what evidence they will look for to know that their seeds are germinating.

ELA Connection: Throughout the module, students use and develop speaking and listening skills through meaningful group discussions, thoughtful explanations, and class presentations. The discussions enable students to share their diverse knowledge and experiences as a whole-class learning opportunity. In addition, students use and develop their reading and writing skills through their STEM Research Notebook entries.

Social Studies Connection: Not applicable.

Activity/Exploration

Mathematics and Science Classes: Students should continue to chart and graph patterns of local weather each day throughout the module to draw connections among weather patterns, seasons, and plant development. Have students water their lima bean and sunflower seeds daily and observe their growth. After seedlings begin to sprout, have students measure the size of their seedlings once a week and record their data on a class chart.

Ask students what they know about why seasons change and what they want to know about the reasons for seasonal changes, documenting student ideas on a KWL chart. Conduct an interactive read-aloud of *Sunshine Makes the Seasons* by Franklyn M. Branley.

STEM Research Notebook Entry #3

After the read-aloud, ask students to reflect on what they learned about the causes of seasonal changes in their STEM Research Notebooks, using both words and pictures.

Terrific Seasonal Tree

Students investigate plant life cycles in this activity. Start by introducing the term *habitat* to the class, asking students for their ideas about what a habitat is. Tell students that a habitat is a place in nature where an animal or plant lives and that many types of animals and plants can share a habitat. Ask students to name some examples of habitats (e.g., a tree is a habitat for animals and insects, a lake is a habitat for fish and plants). Encourage students to think of the town or area where they live as a habitat for people, plants, and animals. Introduce the idea that certain types of plants grow well in your habitat because of the weather (amount of sunlight and rain) and seasonal changes.

Tell students that they are going to look at seasonal changes in a tree that lives in their habitat. Have the class work together to choose a tree that grows in their region (you may wish to limit this to options based on the suggestions in Preparation for Lesson 1 on p. 52). Ask students to predict what part of the tree will reflect the changes in the seasons. Introduce the idea that a tree's *leaves* will change with the seasons. Then, divide the class into four student teams. Assign each team to focus on one season: spring, summer, fall, or winter. Each group will research the effect their assigned season has on the weather in their local habitat as well as on the tree the class picked.

Student teams will research these questions for this activity:

- What is the weather usually like in this season in your habitat?

- How does this season affect the basic needs of the living things in your habitat? (Consider air, water, food, habitat, and sunlight.)

- What happens to your tree during this season?

STEM Research Notebook Entry #4

Have students record their research findings for the above questions in their STEM Research Notebooks, using both words and pictures.

After students have completed their research, each student should create a model of the tree in the season his or her team researched. The focus of this activity is to create a visual representation of students' research. Because these will be presented to the class as a representation of the different seasons, some level of uniformity among the trees will be useful in helping students understand how a tree might change with seasonal changes. This activity therefore includes instructions for creating the tree, rather than having each team or student design the tree differently.

Provide students with the materials for this activity, and direct them to follow these steps:

1. Open the paper bag.

2. From the top opening, cut strips down to the midpoint of the bag.

3. Twist those strips to make branches.

4. Place 2 plastic grocery bag or a few wadded pieces of paper in the lower half of the bag—this will give the tree a base so that it can stand up.

5. Twist the bottom half of the bag to make the trunk.

6. Decorate the tree to represent the season your group is researching. (For example, if it is summer, glue pieces of green paper on the branches to represent green leaves. If it is fall, glue pieces of green, yellow, orange, and red paper on the branches to represent the changing colors of the leaves. If it is winter, glue just a few pieces of brown construction paper on the branches to represent dead leaves. If it is spring, use smaller pieces of green paper to represent buds.)

Student teams should then give short presentations in which they share the results of their research (answers to the questions they researched) and their tree models with the class.

ELA Connection: Begin a class vocabulary chart with new terms introduced during the videos and readings, using both words and pictures. Post the chart on the classroom wall and continue to add to it throughout the module.

STEM Research Notebook Entry #5

Have students add vocabulary words to their STEM Research Notebooks, using both words and pictures.

Social Studies Connection: Not applicable.

Explanation

Mathematics and Science Classes and ELA Connection: Ask students to share what they know and what they want to know about where plants live, recording student ideas on a KWL chart. Then, conduct an interactive read-aloud of *Plants Live Everywhere,* by Mary Dodson Wade, which allows students to explore diversity in plants and plant habitats.

STEM Research Notebook Entry #6

After reading the book, have students document what they learned about where plants live in their STEM Research Notebooks, using both words and pictures.

For Good Measure

Since students will be taking measurements throughout the module, measuring plant heights weekly as well as dimensions for their container garden, this activity provides a review of basic measurement techniques. Show students two objects of differing lengths, such as a pencil and a long piece of string. Ask students to identify which is longer and which is shorter. Ask students to share their ideas about how they would describe the size of each of these objects to someone. Record student ideas on a class chart.

Introduce the idea of measurement as a way to compare sizes of various objects and to express these sizes in a way that people can understand. Hold the pencil vertically and place your hand, with fingers pointed up, next to it. Ask students how tall the pencil is compared with your hand (e.g., a little taller, shorter, about the same). Ask students if they think that saying, "The pencil is a little taller than my hand," is a good way to measure. Have them justify their answers by sharing their reasoning. Then, have students consider the statement "My sister is a little taller than my nephew." Ask students to show, by holding their hands at a certain height, how tall they think the sister is. Ask students if this is a good way to tell someone about size. Lead students to understand that we use special tools to measure items so that everyone can understand sizes without even having to see an object.

Show students a ruler and ask them for their ideas on what it is and how it is used. Tell students that a ruler is one of the special tools that we use to measure items. Give each student a ruler, and review the concept of inches as a unit of measurement. Have students identify the 1 inch mark, 2 inch mark, and so forth on their rulers. Demonstrate to students how to measure the lengths of the pencil and the string to the nearest inch. Then, have students choose an item, such as a crayon or eraser, from their desks to measure. Assist students with measurement, ensuring that they align the end of the ruler with the end of the object being measured, and have them write down the size of the object they measured. Next, have students work in pairs and trade the items they measured. Students should measure their partners' objects and compare measurements. Ask students if their measurements were the same as their partners'. If not, ask students why this might be and have students work together to remeasure the objects.

Social Studies Connection: Ask students what types of plants they see in their neighborhood habitat. Ask students if they have ever seen different types of plants in other parts of the country (possibly when on vacation) that they do not see in their own neighborhood. Help students locate on a map or globe the places in the United States or the world where they have seen different types of plants. Ask students if they think different parts of the United States and the world produce different types of plants and why. Help students identify the various kinds of habitats using images you collected, emphasizing that different types of plants grow in different areas because of different types of weather patterns. The following are examples of various habitats and some of their native plants:

- Arctic tundra: lichen, bearberry

- Desert: cactus, desert lily

- Grassland or prairie: grass, purple coneflower

- Mountain: columbine, wax currant

- Ocean: kelp, seaweed

- Rainforest: orchid, banana

- Wetland: cattails, wild rice

Elaboration/Application of Knowledge

Mathematics and Science Classes and ELA and Social Studies Connections: Remind students of their challenge to create and maintain a container garden. Depending on whether your class is making a window box, free-standing container garden, or raised-bed garden, hold a class discussion about what the term means and why this type of gardening might be used in various areas. Discussion questions might include the following:

- Why would people plant a container garden? (because of limited space, to control use of pesticides, to grow food)

- How would people decide what kinds of plants to put in their container gardens? (different weather conditions, different needs)

Ask students to name different kinds of plants they have seen (e.g., flowers, trees, vegetables, grass). Use a variety of images of different types of landscapes and plants available to prompt students. Tell students that all plants have some parts in common; ask students to name those parts (roots, stems, leaves). Show students the different types of plants you have acquired for the class to observe during the module. Create a class chart for each type of plant, asking students to share their ideas about what observations to include on the charts (e.g., how many leaves, how tall). Working as a class, make observations about each plant and record these on the charts. Ask students to name differences and similarities among the plants. Track the plants' growth weekly as a class through the remainder of the module and over the course of the next several months.

Lesson Assessment. Assess student learning in the lesson by having students do the following:

- Name one plant that lives in their area.

- Identify changes in the plant over the four seasons.

- Draw and label the parts of a plant (roots, stem, leaves).

Evaluation/Assessment

Students may be assessed on the following performance tasks and other measures listed.

Performance Tasks

- Tree models

- Team presentations

- Lesson Assessment

Other Measures (using assessment rubric in Appendix B, p. 134)

- Teacher observations
- STEM Research Notebook entries
- Participation in teams during investigations

INTERNET RESOURCES

STEM Teaching Tools resource
- *http://stemteachingtools.org/pd/sessionc*

Changing Seasons web page
- *www.education.noaa.gov/Climate/Changing_Seasons.html*

The Life Cycle of Flowering Plants web page
- *www.natgeokids.com/za/discover/science/nature/the-life-cycle-of-flowering-plants*

What's the Difference Between Weather and Climate? web page
- *www.nasa.gov/mission_pages/noaa-n/climate/climate_weather.html*

"Trees for All Seasons" interactive poster
- *www.dnr.state.mn.us/forestry/education/treeforallseasons/index.html*

Bureau of Labor Statistics' *Occupational Outlook Handbook*
- *www.bls.gov/ooh/home.htm*

Interactive read-aloud resources
- *www.readingrockets.org/article/repeated-interactive-read-alouds-preschool-and-kindergarten*
- *www.k5chalkbox.com/interactive-read-aloud.html*
- *www.readwritethink.org/professional-development/strategy-guides/teacher-read-aloud-that-30799.html*

Suggested trees by region
- *https://shop.arborday.org/content.aspx?page=tree-nursery*

"The Four Seasons" video by ABCmouse.com
- *www.youtube.com/watch?v=K2tV69N0X8k*

"One Year in 40 Seconds" video
- *www.youtube.com/watch?v=lmIFXIXQQ_E*

SAMPLE WEATHER SYMBOLS FOR CLASS WEATHER CHART

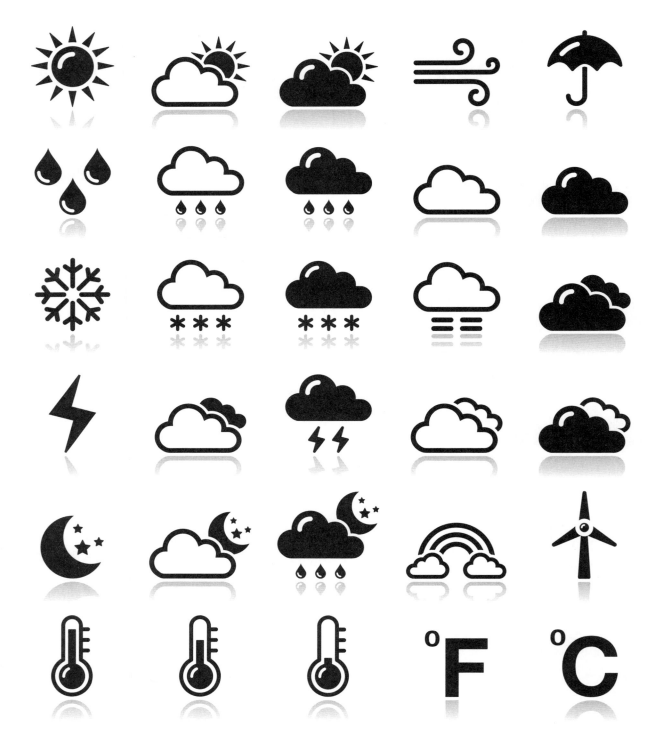

Name: _____

STUDENT HANDOUT

WEATHER CHART: TODAY'S FORECAST

Date	Temperature 	Amount of Sun			Precipitation		Wind		
		Sunny	Partly Sunny	Cloudy	Rain	Snow	Strong Winds	Light Winds	No Wind

Lesson Plan 2: Our Container Garden: Design Time

This lesson introduces students to the engineering design process (EDP), the process they will use to design and create their container garden. In this lesson, students work through the first steps of the EDP to develop a plan for their garden.

ESSENTIAL QUESTIONS

- What are the basic needs of all plants?
- What conditions are necessary for growing plants in a container garden?
- Where do plants come from?
- What types of plants grow from seeds?

ESTABLISHED GOALS AND OBJECTIVES

At the conclusion of this lesson, students will be able to do the following:

- List the basic needs of all plants
- Describe the conditions necessary for growing plants in a container garden
- Explain what happens when all five basic needs are not met
- Identify where plants come from
- Recognize what types of plants grow from seeds
- Use technology to facilitate deeper conceptual understanding about the life cycle of plants
- Identify technological advances and tools that scientists use to learn about the life cycle of plants
- Design a container garden
- Explain concepts through the design of a notebook to make observations of the life cycle of plants

TIME REQUIRED

- 12 days (approximately 30 minutes each day; see Tables 3.7–3.9, pp. 37–38)

MATERIALS

Required Materials for Lesson 2

- STEM Research Notebooks

- Computer with internet access for viewing videos

- Smartphones or tablets for student video recording

- Engineering design process graphic (optional handout per student; attached at the end of this lesson on p. 81)

- Books

  - *Seeds,* by Vijaya Khisty Bodach (Capstone Press, 2016)

  - *Trees, Weeds, and Vegetables: So Many Kinds of Plants!* by Mary Dodson Wade (Enslow Publishers, 2009)

  - *Plant Secrets,* by Emily Goodman (Charlesbridge, 2009)

  - *Flower Garden,* by Eve Bunting (Voyager Books, 2000)

- Chart paper

- Rulers (1 per student)

- Pencils (1 per student)

- Crayons for use in STEM Research Notebook entries (1 set per student)

CONTENT STANDARDS AND KEY VOCABULARY

Table 4.4 lists the content standards from the *NGSS, CCSS,* NAEYC, and the Framework for 21st Century Learning that this lesson addresses, and Table 4.5 (p. 69) presents the key vocabulary. Vocabulary terms are provided for both teacher and student use. Teachers may choose to introduce some or all of the terms to students.

Table 4.4. Content Standards Addressed in STEM Road Map Module Lesson 2

NEXT GENERATION SCIENCE STANDARDS
PERFORMANCE EXPECTATIONS • 1-ESS1-1. Use observations of the sun, moon, and stars to describe patterns that can be predicted. • 1-ESS1-2. Make observations at different times of year to relate the amount of daylight to the time of year.

Continued

Table 4.4. (*continued*)

SCIENCE AND ENGINEERING PRACTICES

Planning and Carrying Out Investigations

Planning and carrying out investigations to answer questions or test solutions to problems in K–2 builds on prior experiences and progresses to simple investigations, based on fair tests, which provide data to support explanations or design solutions.

• Plan and conduct investigations collaboratively to produce evidence to answer a question.

Analyzing and Interpreting Data

Analyzing data in K–2 builds on prior experiences and progresses to collecting, recording, and sharing observations.

• Use observations (firsthand or from media) to describe patterns in the natural world in order to answer scientific questions.

Constructing Explanations and Designing Solutions

Constructing explanations and designing solutions in K–2 builds on prior experiences and progresses to the use of evidence and ideas in constructing evidence-based accounts of natural phenomena and designing solutions.

• Make observations (firsthand or from media) to construct an evidence-based account for natural phenomena.

Obtaining, Evaluating, and Communicating Information

Obtaining, evaluating, and communicating information in K–2 builds on prior experiences and uses observations and texts to communicate new information.

• Read grade-appropriate texts and use media to obtain scientific information to determine patterns in the natural world.

DISCIPLINARY CORE IDEAS

ESS1.A: The Universe and Its Stars

• Patterns of the motion of the sun, moon, and stars in the sky can be observed, described, and predicted.

ESS1.B: Earth and the Solar System

• Seasonal patterns of sunrise and sunset can be observed, described, and predicted.

LS1.A: Structure and Function

• All organisms have external parts. Different animals use their body parts in different ways to see, hear, grasp objects, protect themselves, move from place to place, and seek, find, and take in food, water and air. Plants also have different parts (roots, stems, leaves, flowers, fruits) that help them survive and grow.

LS1.B: Growth and Development of Organisms

• Adult plants and animals can have young. In many kinds of animals, parents and the offspring themselves engage in behaviors that help the offspring to survive.

Continued

Table 4.4. (*continued*)

LS1.D: Information Processing

- Animals have body parts that capture and convey different kinds of information needed for growth and survival. Animals respond to these inputs with behaviors that help them survive. Plants also respond to some external inputs.

CROSSCUTTING CONCEPTS

Patterns

- Patterns in the natural world can be observed, used to describe phenomena, and used as evidence.

Structure and Function

- The shape and stability of structures of natural and designed objects are related to their function(s).

COMMON CORE STATE STANDARDS FOR MATHEMATICS

MATHEMATICAL PRACTICES

- MP1. Make sense of problems and persevere in solving them.
- MP2. Reason abstractly and quantitatively.
- MP3. Construct viable arguments and critique the reasoning of others.
- MP4. Model with mathematics.
- MP5. Use appropriate tools strategically.
- MP6. Attend to precision.
- MP7. Look for and make use of structure.
- MP8. Look for and express regularity in repeated reasoning.

MATHEMATICAL CONTENT

- 1.NBT.B.3. Compare two two-digit numbers based on meanings of the tens and ones digits, recording the results of comparisons with the symbols >, =, and <.
- 1.MD.A.1. Order three objects by length; compare the lengths of two objects indirectly by using a third object.
- 1.MD.C.4. Organize, represent, and interpret data with up to three categories; ask and answer questions about the total number of data points, how many in each category, and how many more or less are in one category than in another.
- 1.OA.A.1. Use addition and subtraction within 20 to solve word problems involving situations of adding to, taking from, putting together, taking apart, and comparing, with unknowns in all positions.

Continued

Table 4.4. (*continued*)

> • 1.OA.A.2. Solve word problems that call for addition of three whole numbers whose sum is less than or equal to 20, e.g., by using objects, drawings, and equations with a symbol for the unknown number to represent the problem.
>
> ## COMMON CORE STATE STANDARDS FOR ENGLISH LANGUAGE ARTS
>
> ### READING STANDARDS
>
> • RI.1.1. Ask and answer questions about key details in a text.
>
> • RI.1.2. Identify the main topic and retell key details of a text.
>
> • RI.1.3. Describe the connection between two individuals, events, ideas, or pieces of information in a text.
>
> • RI.1.7. Use the illustrations and details in a text to describe its key ideas.
>
> ### WRITING STANDARDS
>
> • W.1.2. Write informative/explanatory texts in which they name a topic, supply some facts about the topic, and provide some sense of closure.
>
> • W.1.6. With guidance and support from adults, use a variety of digital tools to produce and publish writing, including in collaboration with peers.
>
> • W.1.7. Participate in shared research and writing.
>
> • W.1.8. With guidance and support from adults, recall information from experiences or gather information from provided sources to answer a question.
>
> ### SPEAKING AND LISTENING STANDARDS
>
> • SL.1.1. Participate in collaborative conversations with diverse partners about *grade 1 topics and texts* with peers and adults in small and larger groups.
>
> • SL.1.1.A. Follow agreed-upon rules for discussions.
>
> • SL.1.1.B. Build on others' talk in conversations by responding to the comments of others through multiple exchanges.
>
> • SL.1.1.C. Ask questions to clear up any confusion about the topics and texts under discussion.
>
> • SL.1.3. Ask and answer questions about what a speaker says in order to gather additional information or clarify something that is not understood.
>
> • SL.1.5. Add drawings or other visual displays to descriptions when appropriate to clarify ideas, thoughts, and feelings.
>
> ## NATIONAL ASSOCIATION FOR THE EDUCATION OF YOUNG CHILDREN STANDARDS
>
> • 2.E.1. Arrange firsthand, meaningful experiences that are intellectually and creatively stimulating, invite exploration and investigation, and engage children's active, sustained involvement by providing a rich variety of material, challenges, and ideas.

Continued

Table 4.4. (*continued*)

- 2.F.3. Extend the range of children's interests and the scope of their thought, present novel experiences and introduce stimulating ideas, problems, experiences, or hypotheses.

- 2.F.6. Enhance children's conceptual understanding through various strategies, including intensive interview and conversation, encourage children to reflect on and "revisit" their experiences.

- 2.G.2. Scaffolding takes on a variety of forms.

- 2.J.1. Incorporate a wide variety of experiences, materials and equipment, and teaching strategies to accommodate the range of children's individual differences in development, skills and abilities, prior experiences, needs, and interests.

- 3.A.1. Teachers consider what children should know, understand, and be able to do across the domains.

FRAMEWORK FOR 21ST CENTURY LEARNING
- Interdisciplinary Themes; Learning and Innovation Skills; Information, Media, and Technology Skills; Life and Career Skills

Table 4.5. Key Vocabulary for Lesson 2

Key Vocabulary	Definition
herb	a plant that is used to flavor food
hydroponics	a way of growing plants without soil that uses water to deliver nutrients to plants
philanthropy	the desire or effort to do good things for other people, such as an act or gift to benefit others, often for the common good
repellent	something that keeps insects or other pests away
vertical farming	a way of growing plants using containers stacked in layers often used where there is little growing space or where natural growing conditions for plants are challenging

TEACHER BACKGROUND INFORMATION

Students create designs for their container garden in this lesson. The procedures students use to plant and maintain the garden will vary according to the type of garden (window box, free-standing container, or raised-bed garden), the time of year, and the type of plants included (flowers, herbs, or vegetables). Students will conduct research and make decisions about their garden as they proceed through their planning in this lesson. Their container garden may be as simple as a series of milk jugs and 2 liter bottles

used as planters or as complex as an outdoor raised bed framed with wood. You may find the following online resources useful:

- "How to Grow a Window Sill Garden" article: *www.wikihow.com/Grow-a-Window-Sill-Garden*

- Designing a School Garden web page: *www.kidsgardening.org/designing-a-school-garden*

- "How to Plant Window Boxes—Gardening Tutorial" video: *www.youtube.com/watch?v=7OfG1zOKuMA*

- Recycled Plastic Container Planters to Make With Kids web page: *www.reasonstoskipthehousework.com/recycled-plastic-container-planters/recycle-planters-8*

Engineering

Students begin to gain an understanding of engineering as a profession in this lesson as they learn to use the EDP to create a plan for their container garden. Students should understand that engineers are people who design and build products and systems in response to human needs. For an overview of the various types of engineering professions, see the following websites:

- *www.engineergirl.org/33/TryOnACareer*

- *www.nacme.org/types-of-engineering*

- *www.sciencekids.co.nz/sciencefacts/engineering/typesofengineeringjobs.html*

Engineering Design Process

Students should understand that engineers need to work in groups to accomplish their work and that collaboration is important for designing solutions to problems. In this lesson and the next one, students will use the EDP, the same process that professional engineers use in their work. A graphic representation of the EDP is provided at the end of this lesson. You may wish to provide each student with a copy of the EDP graphic or enlarge it and post it in a prominent place in your classroom for student reference throughout the module. Be prepared to review each step of the EDP listed on the graphic with students and emphasize that the process is not a linear one—at any point in the process, they may need to return to a previous step. The steps of the process are as follows:

1. *Define.* Describe the problem you are trying to solve, identify what materials you can use, and determine how much time and help you have to solve the problem.

2. *Learn.* Brainstorm solutions and conduct research to learn about the problem you are trying to solve.

3. *Plan.* Plan your work, including making sketches and dividing tasks among team members if necessary.

4. *Try.* Build a device, create a system, or complete a product.

5. *Test.* Now, test your solution. This might be done by conducting a performance test, if you have created a device to accomplish a task, or by asking for feedback from others about their solutions to the same problem.

6. *Decide.* Based on what you found out during the Test phase, you can adjust your solution or make changes to your device.

After completing all six steps, students can share their solutions or devices with others. This represents an additional opportunity to receive feedback and make additional modifications based on that feedback.

In this lesson, students work as a class to proceed through the first three steps of the EDP—Define, Learn, and Plan—as they address the module challenge. The following are additional resources about the EDP:

- *www.sciencebuddies.org/engineering-design-process/engineering-design-compare-scientific-method.shtml*

- *www.pbslearningmedia.org/resource/phy03.sci.engin.design.desprocess/what-is-the-design-process*

COMMON MISCONCEPTIONS

Students will have various types of prior knowledge about the concepts introduced in this lesson. Table 4.6 (p. 72) outlines some common misconceptions students may have concerning these concepts. Because of the breadth of students' experiences, it is not possible to anticipate every misconception that students may bring as they approach this lesson. Incorrect or inaccurate prior understanding of concepts can influence student learning in the future, however, so it is important to be alert to misconceptions such as those presented in the table.

Table 4.6. Common Misconceptions About the Concepts in Lesson 2

Topic	Student Misconception	Explanation
Engineers and the engineering design process (EDP)	All engineers are people who drive trains.	Railroad engineers are just one type of engineer. The engineers referred to in this module are people who use science, technology, and mathematics to build machines, products, and structures that meet people's needs.
	Engineers use only science and mathematics to do their work.	Engineers often use science and mathematics in their work, but they also use many other kinds of knowledge to solve problems and design products, such as how people use products, what people's needs are, and how the natural environment affects materials.
	Engineers work alone to build things.	Engineers often work in teams and use a process to solve problems. The process involves creative thinking, research, and planning, in addition to building and testing products.
Living things and resources	Different kinds of organisms do not compete with each other for resources such as water.	All living things need food, water, shelter, sunlight, and air and compete with each other for these resources.
Plants	Plants take in everything they need to grow through their roots.	Plants take in water and minerals through their roots; however, they also take in air and absorb sunlight through their leaves.
	Sunlight helps plants grow by keeping them warm.	Sunlight is absorbed by chloroplasts for use in photosynthesis.

PREPARATION FOR LESSON 2

Review the Teacher Background Information provided, assemble the materials for the lesson, duplicate the EDP graphic if you wish to hand this out to students or enlarge it to post in the classroom, and preview the videos recommended in the Learning Components section that follows.

Prepare resources in advance or plan a trip to the school media center for student teams to conduct research during the Learn phase of the EDP during this lesson. You might also enlist adult volunteers or older students to help your students as they conduct research. You may wish to invite a guest speaker from a local garden center or a horticulturist or botanist to speak with the class about options for plants as part of their research; have students prepare questions in advance. Having a guest speaker who works with plants may also help students connect their understanding of plant development with specific careers.

LEARNING COMPONENTS
Introductory Activity/Engagement

Connection to the Challenge: Begin each day of this lesson by directing students' attention to the module challenge, the Container Garden Design Challenge:

> *Your town's leaders have noticed that many of the fruits, vegetables, and flowers sold in your grocery stores were grown far away from your local community. They think that growing more of these items locally will provide produce that is fresher and less expensive. Because of this, the town's leaders are looking for ways for families, schools, and businesses to use space creatively to grow fruits, vegetables, and flowers for members of the community to use. Your class has been challenged to create a garden to grow produce or other plants at your school as a model for the community of how this can be done and how your local weather patterns affect plants.*

Hold a brief class discussion of how students' learning in the previous days' lessons contributed to their ability to complete the challenge. You may wish to create a class list of key ideas on chart paper.

Mathematics and Science Classes: Introduce the concept that living things all have basic needs. Following agreed-upon rules for discussions, ask students the following questions and then document student responses on a KWL chart:

- What are the basic needs of all living things? (air, water, food, habitat, and sunlight)

- What conditions are necessary for growing plants in a container garden? (the five basic needs)

- What happens when all five basic needs are not met?

- Where do plants come from?

- What types of plants grow from seeds? (e.g., fruit, flowers, herbs, grass, trees)

Ask students to share what they know and what they want to know about seeds, documenting student ideas on a KWL chart. Then, conduct an interactive read-aloud

of *Seeds,* by Vijaya Khisty Bodach, through which students will learn about plants that come from seeds.

STEM Research Notebook Entry #7

After the read-aloud, have students document what they learned about seeds in their STEM Research Notebooks, using both words and pictures.

Then, tell students that as part of their module challenge, they need to decide what types of plants to include in their container garden. Ask students what they know and want to know about different types of plants, documenting student ideas on a KWL chart. Conduct an interactive read-aloud of *Trees, Weeds, and Vegetables: So Many Kinds of Plants!* by Mary Dodson Wade, which will allow students to explore various plants.

STEM Research Notebook Entry #8

After the read-aloud, have students document what they learned about types of plants in their STEM Research Notebooks, using both words and pictures.

ELA Connection: Not applicable.

Social Studies Connection: Not applicable.

Activity/Exploration

Mathematics and Science Classes and ELA Connection: Introduce students to the EDP and the idea that engineers and other workers who design things use a process to achieve their goals. Explain each of the steps of the EDP (see Teacher Background Information on p. 69 and the EDP graphic attached at the end of this lesson on p. 81).

Direct students' attention to the Define phase of the EDP. Ask students what problem they are being asked to solve in their challenge (how to grow plants in small spaces). Using student responses, create a definition of the problem in sentence form. Record this definition where students can refer to it during the remainder of the module. Ask students what things they need to know to design a garden and choose plants (the weather in your area, what kind of plants will grow at this time of year, plants' needs, what space they can use at the school, and so on). Refer to the Learn phase of the EDP and tell students that they will learn about these things by conducting research.

Ask students to share what they know and want to know about container gardening, documenting student ideas on a KWL chart. Show a video about container gardening such as "Container Gardening With Kids" at *www.youtube.com/watch?v=zgK6BNKArS0.* This time, allow for student involvement during the video as described for interactive read-alouds in the Teacher Background Information section for Lesson 1 (see p. 50).

STEM Research Notebook Entry #9

After viewing the video, ask students to document what they learned about container gardening in their STEM Research Notebooks, using both words and pictures.

To complete the Learn phase of the EDP, assign each team a topic to research. Teams will then share their findings with the class. Assign each team one of the following questions to research using the resources you have assembled or at the school media center:

1. Where can you put your garden? (Is there a space and a way to attach a window box? Is there a place outside for a raised bed or free-standing container garden? How much space do you have in the classroom for a garden? How much space is required for each sort of garden: window box, free-standing container, and raised bed?)

2. What can you plant in your garden? (What time of year is it? What is the weather like at this time of year where you live? What kind of plants grow where you live?)

3. What materials do you need to make a window box, free-standing container garden, or raised bed?

4. How will you control pests? (What kind of homemade insect repellents can you use? If your garden is outside, how can you keep animals away from it?)

The students on each team should first brainstorm their ideas about the assigned question and then conduct their research. Students should record their findings using pictures and words. After completing their research, have students present their team's findings to the class by reading their question, giving their team's answer to the question, and displaying their pictures. An option is for you to create a slide presentation of students' completed research findings (pictures and text) for teams to use in their presentations to the class.

After student teams have presented their information, tell the class that now they have moved through the first two phases of the EDP: Define and Learn. They are now ready to move into the Plan phase. During this phase, class members should make final decisions about the type of garden they will create (window box, free-standing container, or raised bed), its location, the types of plants they will include, pest control, and what resources they will need to construct the garden. The class should make decisions about each of these issues using the teams' research as a guide for decision making. Once the class has decided on a type of garden, its location, and what they will plant, create a class list of materials necessary to create the garden. Be sure to consider how much soil will be needed and how many plants the garden can sustain.

STEM Research Notebook Entry #10

Ask students to create diagrams of the container garden design. These drawings should include and label location, materials (e.g., wood, recycled milk jugs, recycled bottles and cans), and measures for pest control and watering. Students should be mindful of issues such as providing adequate space for each plant as they create their designs.

Then, have students share their design drawings with the class. The class as a whole should choose one design, or combine features of several designs, to use to create the garden. Make a working drawing on chart paper for class reference as students build the garden; this may be a larger version of one student's design or a combination of several students' designs. Label all materials and note measurements on the class drawing.

Next, ask students to share their ideas about plant life cycles, recording what students know and what they want to know on a KWL chart. Conduct an interactive read-aloud of *Plant Secrets*, by Emily Goodman, about plant growth cycles, asking students to listen for new vocabulary terms.

STEM Research Notebook Entry #11

After the read-aloud, have students document what they learned about plant growth in their STEM Research Notebooks, using both words and pictures.

Social Studies Connection: Discuss with students the impact gardening has on culture. For example, growing traditional plants from one's native culture can promote a sense of cultural identity and cultural stewardship. Ask students what foods are special in their families and to their cultural heritage (for example, special holiday dishes). Ask students if their families have a garden, and if so, what foods they plant and how they use them in family dishes.

Explanation

Mathematics and Science Classes: After tracking the weather for at least two weeks, you can begin to analyze the data for trends (exactly how and when you do this will depend on where you live and what you have decided to track on the chart). In general, analysis of the weather data should be focused on identifying patterns. To do this, display all the data and ask for student ideas about how to organize this data to see if there are any patterns. Have students make suggestions (e.g., count the number of rainy days and sunny days) and guide students to identify categories (e.g., sunny days, rainy days, cloudy days). Create a class chart of these categories. You can then have students create a bar graph of their findings, adding to it each week or each day, depending on the categories the class chose.

Students will identify and list supply needs for their container garden in their next STEM Research Notebook entry as part of the Plan phase of the EDP. Prompt students by asking the following questions:

- What supplies do we need to make a container garden planter? (e.g., wood, recycled milk jugs, recycled bottles and cans, art supplies)

- How much soil do we need?

- What seeds or seedlings do we need?

- How many seeds or seedlings do we need?

- What tools do we need?

- How can we provide enough water?

Optional: You may wish to have students design a watering system using recycled plastic bottles and jugs. The watering system may be as simple as a series of plastic bottles and directions for how often to water the garden, or students might propose a schedule for watering the garden using the bottles. The plan could also include labeling the bottles with their purpose and planning where to store them so that they do not blow away outdoors. Alternatively, students may plan to harvest rainwater in the bottles, in which case they would need to remove the tops and devise a method for keeping the bottles from blowing away (e.g., putting pebbles in the bottom). The point of this optional activity is for students to think creatively and in detail about plants' water needs.

STEM Research Notebook Entry #12

Tell students to identify, using words and pictures, the supplies they need to start and maintain their container garden.

Ask students to review the list of materials they created and offer ideas about how those materials could be supplied (e.g., students could bring in empty milk jugs from home, families might donate scrap wood). Record student responses on a class list.

STEM Research Notebook Entry #13

Have students review their list of supply needs and record their ideas about how they can contribute some of these supply needs for the garden.

ELA and Social Studies Connections: Conduct an interactive read-aloud of *Flower Garden,* by Eve Bunting, through which students will make a literature connection to their supply needs for their container garden.

STEM Research Notebook Entry #14

After the read-aloud, have students document their reactions to the story and what they learned about purchasing garden supplies in their STEM Research Notebooks, using both words and pictures.

Elaboration/Application of Knowledge

Mathematics and Science Classes: Introduce the concept of urban farming by showing a video such as "Grow—Episode 6: Thrive" from Whole Foods Market at *www.youtube.com/watch?v=Rr3Ercz5bUI*. The video shown should allow students to observe what materials are used in creating the garden. Ask students to watch for what materials are needed for the garden. Create a class list of students' ideas and compare with the list of materials students have created for their garden.

Revisit the supply list students created for their garden and ask students for ideas of items they could recycle or bring from home to supply some of these needs (e.g., empty milk jugs, wooden boxes).

STEM Research Notebook Entry #15

Have students create a list of items that could be recycled or brought from home to meet supply needs for the garden.

Lesson Assessment. Assess student learning in the lesson by having students do the following:

- List the five basic needs of all plants.

- Describe two conditions necessary for growing plants in a container garden.

- Explain what happens when all five basic needs are not met.

- Name three types of plants that grow from seeds.

ELA Connection: Not applicable.

Social Studies Connection: Ask students to share their ideas about what they should do with the fruits, vegetables, herbs, or flowers that their container garden produces (e.g., donate these to the school cafeteria or a local food bank, take produce or flowers home to their families).

Discuss with students the idea of philanthropy, the desire to do good for other people, such as an act or gift to benefit others, often done for the common good. Philanthropy through donation of vegetables, herbs, and fruits to the school cafeteria or a food bank provides the opportunity to teach students that teamwork and personal responsibility are important and valuable assets to the school and to the community at large. This experience will enhance students' self-esteem while teaching problem-solving abilities and social skills.

Continue to discuss the cultural and societal implications of container gardens. For example, container gardening allows for the consumption of healthy food grown without the use of pesticides and herbicides. Therefore, container gardening is an excellent way to grow rich, nutritious food in small spaces.

Help students connect their understanding of plant development by discussing specific plant- or weather-related careers (see the list of related careers on p. 49). You may wish to invite a guest speaker from one of these professions. Identify technological advances that are related to plants and plant development, such as hydroponics and vertical farming. Optional: You might want to plan a class field trip to a garden center or local botanical garden to learn about careers related to growing plants and about technological advances related to horticulture.

Evaluation/Assessment

Students may be assessed on the following performance tasks and other measures listed.

Performance Tasks

- Team research and presentation for Container Garden Design Challenge

- Lesson Assessment

Other Measures (using assessment rubric in Appendix B, p. 134)

- Teacher observations

- STEM Research Notebook entries

- Participation in teams during investigations

INTERNET RESOURCES

"How to Grow a Window Sill Garden" article
- *www.wikihow.com/Grow-a-Window-Sill-Garden*

Designing a School Garden web page
- *www.kidsgardening.org/designing-a-school-garden*

"How to Plant Window Boxes—Gardening Tutorial" video
- *www.youtube.com/watch?v=7OfG1zOKuMA*

Recycled Plastic Container Planters to Make With Kids web page
- *www.reasonstoskipthehousework.com/recycled-plastic-container-planters/recycle-planters-8*

Overview of the various types of engineering professions
- *www.engineergirl.org/33/TryOnACareer*
- *www.nacme.org/types-of-engineering*
- *www.sciencekids.co.nz/sciencefacts/engineering/typesofengineeringjobs.html*

Additional resources about the EDP
- *www.sciencebuddies.org/engineering-design-process/engineering-design-compare-scientific-method.shtml*
- *www.pbslearningmedia.org/resource/phy03.sci.engin.design.desprocess/what-is-the-design-process*

"Container Gardening With Kids" video
- *www.youtube.com/watch?v=zgK6BNKArS0*

"Grow—Episode 6: Thrive" video from Whole Foods Market
- *www.youtube.com/watch?v=Rr3Ercz5bUI*

ENGINEERING DESIGN PROCESS

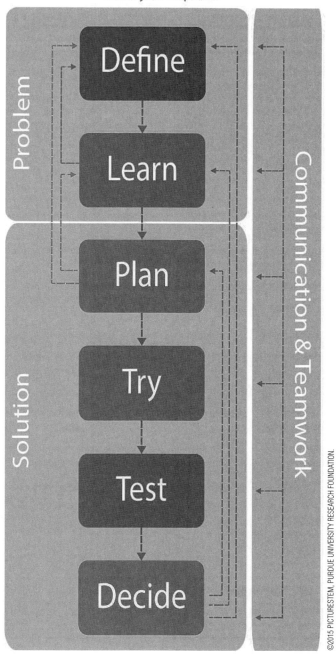

Lesson Plan 3: Our Container Garden: Planting Time

In this lesson, students continue to use the EDP to create the class container garden. Students move into the Try phase as they construct the garden. Through the construction and planting of their container garden and subsequent observations, students begin to develop conceptual awareness of how living things grow and change. Students create an observation journal in which to track garden data on an ongoing basis. The Test and Improve phases of the EDP will be executed during the remainder of the school year as the class continues to observe and care for the garden.

ESSENTIAL QUESTIONS

- What do your seeds need to sprout into seedlings?
- How long will it take for your seeds to sprout into seedlings?
- How much will your seedlings grow each week in your container garden?
- How can we record data about our garden?

ESTABLISHED GOALS AND OBJECTIVES

At the conclusion of this lesson, students will be able to do the following:

- Determine what seeds need to sprout into seedlings
- Estimate how much time it will take for seeds to sprout into seedlings
- Estimate how much seedlings will grow each week
- Demonstrate conceptual awareness of about how living things grow and change over the course of their lives by understanding that plants change over time due to the changing seasons
- Construct a container garden
- Observe, measure, quantify, and analyze data during the life cycle of plants

TIME REQUIRED

- 6 days (approximately 30 minutes each day; see Tables 3.9–3.10, pp. 38–39)

MATERIALS

Required Materials for Lesson 3

- STEM Research Notebooks

- Computer with internet access for viewing videos

- Smartphones or tablets for student video recording

- Books

  - *From Seed to Plant,* by Gail Gibbons (Holiday House, 1991)

  - *How a Seed Grows,* by Helene J. Jordan (HarperCollins, 1992)

  - *Plants!* by Brenda Iasevoli (HarperCollins, 2006)

- Chart paper

- Measuring tape or yardstick

- Rulers (1 per student)

- Container garden supplies (may vary depending on students' design):

  - Wood

  - Recycled milk jugs

  - Recycled bottles and cans

  - Potting soil

  - Seeds or seedlings

- Clipboards for students to use in outdoor plant observations

- Pencils (1 per student)

- Crayons or colored pencils

- Safety glasses or goggles, nonlatex aprons, and nonlatex gloves

SAFETY NOTES

1. Remind students that personal protective equipment (safety glasses or goggles, aprons, and gloves) must be worn during all phases of this inquiry activity.

2. Caution students not to eat seeds, which may be coated with toxic chemicals such as herbicides, pesticides, or fungicides.

3. Immediately wipe up any spilled water or soil on the floor to avoid a slip-and-fall hazard.

4. Tell students to be careful when handling recycled bottles and cans. Cans may have sharp edges, which can cut or puncture skin. Glass or plastic bottles can break and cut skin.

5. Have students wash hands with soap and water after the activity is completed.

CONTENT STANDARDS AND KEY VOCABULARY

Table 4.7 lists the content standards from the *NGSS, CCSS,* NAEYC, and the Framework for 21st Century Learning that this lesson addresses, and Table 4.8 (p. 88) presents the key vocabulary. Vocabulary terms are provided for both teacher and student use. Teachers may choose to introduce some or all of the terms to students.

Table 4.7. Content Standards Addressed in STEM Road Map Module Lesson 3

NEXT GENERATION SCIENCE STANDARDS

PERFORMANCE EXPECTATIONS

- 1-ESS1-1. Use observations of the sun, moon, and stars to describe patterns that can be predicted.

- 1-ESS1-2. Make observations at different times of year to relate the amount of daylight to the time of year.

- 1-LS1-2. Read texts and use media to determine patterns in behavior of parents and offspring that help offspring survive.

- 1-LS3-1. Make observations to construct an evidence-based account that young plants and animals are like, but not exactly like, their parents.

SCIENCE AND ENGINEERING PRACTICES

Planning and Carrying Out Investigations

Planning and carrying out investigations to answer questions or test solutions to problems in K–2 builds on prior experiences and progresses to simple investigations, based on fair tests, which provide data to support explanations or design solutions.

- Plan and conduct investigations collaboratively to produce evidence to answer a question.

Analyzing and Interpreting Data

Analyzing data in K–2 builds on prior experiences and progresses to collecting, recording, and sharing observations.

- Use observations (firsthand or from media) to describe patterns in the natural world in order to answer scientific questions.

Continued

Table 4.7. (*continued*)

Constructing Explanations and Designing Solutions

Constructing explanations and designing solutions in K–2 builds on prior experiences and progresses to the use of evidence and ideas in constructing evidence-based accounts of natural phenomena and designing solutions.

- Make observations (firsthand or from media) to construct an evidence-based account for natural phenomena.

Obtaining, Evaluating, and Communicating Information

Obtaining, evaluating, and communicating information in K–2 builds on prior experiences and uses observations and texts to communicate new information.

- Read grade-appropriate texts and use media to obtain scientific information to determine patterns in the natural world.

DISCIPLINARY CORE IDEAS

ESS1.A: The Universe and Its Stars

- Patterns of the motion of the sun, moon, and stars in the sky can be observed, described, and predicted.

ESS1.B: Earth and the Solar System

- Seasonal patterns of sunrise and sunset can be observed, described, and predicted.

LS1.A: Structure and Function

- All organisms have external parts. Different animals use their body parts in different ways to see, hear, grasp objects, protect themselves, move from place to place, and seek, find, and take in food, water and air. Plants also have different parts (roots, stems, leaves, flowers, fruits) that help them survive and grow.

LS1.B: Growth and Development of Organisms

- Adult plants and animals can have young. In many kinds of animals, parents and the offspring themselves engage in behaviors that help the offspring to survive.

LS1.D: Information Processing

- Animals have body parts that capture and convey different kinds of information needed for growth and survival. Animals respond to these inputs with behaviors that help them survive. Plants also respond to some external inputs.

LS3.A: Inheritance of Traits

- Young animals are very much, but not exactly, like their parents. Plants also are very much, but not exactly, like their parents.

LS3.B: Variation of Traits

- Individuals of the same kind of plant or animal are recognizable as similar but can also vary in many ways.

Continued

Table 4.7. (*continued*)

CROSSCUTTING CONCEPTS

Patterns
- Patterns in the natural world can be observed, used to describe phenomena, and used as evidence.

Structure and Function
- The shape and stability of structures of natural and designed objects are related to their function(s).

COMMON CORE STATE STANDARDS FOR MATHEMATICS

MATHEMATICAL PRACTICES
- MP1. Make sense of problems and persevere in solving them.
- MP2. Reason abstractly and quantitatively.
- MP3. Construct viable arguments and critique the reasoning of others.
- MP4. Model with mathematics.
- MP5. Use appropriate tools strategically.
- MP6. Attend to precision.
- MP7. Look for and make use of structure.
- MP8. Look for and express regularity in repeated reasoning.

MATHEMATICAL CONTENT
- 1.NBT.B.3. Compare two two-digit numbers based on meanings of the tens and ones digits, recording the results of comparisons with the symbols >, =, and <.
- 1.MD.A.1. Order three objects by length; compare the lengths of two objects indirectly by using a third object.
- 1.MD.C.4. Organize, represent, and interpret data with up to three categories; ask and answer questions about the total number of data points, how many in each category, and how many more or less are in one category than in another.
- 1.OA.A.1. Use addition and subtraction within 20 to solve word problems involving situations of adding to, taking from, putting together, taking apart, and comparing, with unknowns in all positions.
- 1.OA.A.2. Solve word problems that call for addition of three whole numbers whose sum is less than or equal to 20, e.g., by using objects, drawings, and equations with a symbol for the unknown number to represent the problem.

Continued

Table 4.7. (*continued*)

COMMON CORE STATE STANDARDS FOR ENGLISH LANGUAGE ARTS

READING STANDARDS

- RI.1.1. Ask and answer questions about key details in a text.

- RI.1.2. Identify the main topic and retell key details of a text.

- RI.1.3. Describe the connection between two individuals, events, ideas, or pieces of information in a text.

- RI.1.7. Use the illustrations and details in a text to describe its key ideas.

WRITING STANDARDS

- W.1.2. Write informative/explanatory texts in which they name a topic, supply some facts about the topic, and provide some sense of closure.

- W.1.6. With guidance and support from adults, use a variety of digital tools to produce and publish writing, including in collaboration with peers.

- W.1.7. Participate in shared research and writing.

- W.1.8. With guidance and support from adults, recall information from experiences or gather information from provided sources to answer a question.

SPEAKING AND LISTENING STANDARDS

- SL.1.1. Participate in collaborative conversations with diverse partners about *grade 1 topics and texts* with peers and adults in small and larger groups.

- SL.1.1.A. Follow agreed-upon rules for discussions.

- SL.1.1.B. Build on others' talk in conversations by responding to the comments of others through multiple exchanges.

- SL.1.1.C. Ask questions to clear up any confusion about the topics and texts under discussion.

- SL.1.3. Ask and answer questions about what a speaker says in order to gather additional information or clarify something that is not understood.

- SL.1.5. Add drawings or other visual displays to descriptions when appropriate to clarify ideas, thoughts, and feelings.

NATIONAL ASSOCIATION FOR THE EDUCATION OF YOUNG CHILDREN STANDARDS

- 2.E.1. Arrange firsthand, meaningful experiences that are intellectually and creatively stimulating, invite exploration and investigation, and engage children's active, sustained involvement by providing a rich variety of material, challenges, and ideas.

- 2.F.3. Extend the range of children's interests and the scope of their thought, present novel experiences and introduce stimulating ideas, problems, experiences, or hypotheses.

- 2.F.6. Enhance children's conceptual understanding through various strategies, including intensive interview and conversation, encourage children to reflect on and "revisit" their experiences.

Continued

Table 4.7. (*continued*)

> - 2.G.2. Scaffolding takes on a variety of forms.
>
> - 2.J.1. Incorporate a wide variety of experiences, materials and equipment, and teaching strategies to accommodate the range of children's individual differences in development, skills and abilities, prior experiences, needs, and interests.
>
> - 3.A.1. Teachers consider what children should know, understand, and be able to do across the domains.
>
> **FRAMEWORK FOR 21ST CENTURY LEARNING**
> - Interdisciplinary Themes; Learning and Innovation Skills; Information, Media, and Technology Skills; Life and Career Skills

Table 4.8. Key Vocabulary for Lesson 3

Key Vocabulary	Definition
drainage	the removal of excess water to ensure that plants stay healthy
fertilizer	a substance added to soil to provide nutrients to plants to help them grow
insect repellent	a substance applied to a surface that insects will avoid; discourages insects from landing on a surface

TEACHER BACKGROUND INFORMATION
Plant Growth and Garden Maintenance

In this lesson, students will create the container garden and a plan to maintain the garden. In these processes, you will prompt students to consider issues such as drainage, fertilizing, and pest control.

Drainage is important in container gardens since water accumulation due to over-watering or rain could result in root rot, a condition in which the roots of the plant rot due to fungi in the soil. Plants with root rot are unable to absorb nutrients from the soil and exhibit stunted growth and wilting leaves and will eventually die. To prevent root rot, ensure that there are drainage holes at the bottom of the containers used in the container garden and that water can flow freely away from these holes. For more information on drainage and root rot, see the University of Illinois Extension's Successful Container Gardens: Drainage web page at *www.extension.illinois.edu/containergardening/choosing_drainage.cfm.*

When gardening in containers, the naturally occurring nutrients in soil will be depleted over time and can also be leached from the soil by frequent watering. This means that adding fertilizers is necessary to restore nutrients to the soil. Many types of fertilizers are available, including time-release and liquid fertilizers. The type of fertilizer you use is dependent on the plants in the container garden. Your local garden center should be able to provide you with guidance about appropriate fertilizers for the plants in the class container garden. For more information about fertilizers, see the University of Illinois Extension's Successful Container Gardens: Fertilizers web page at *www.extension. illinois.edu/containergardening/fertilizing.cfm*.

Pest control is also a consideration with container gardens. When insects are visible on plants, they can be hand-picked off the plant (be sure to follow safety guidelines, including wearing plastic gloves, if students remove insects from the plants). Other options for pest control include non-toxic insect repellents that can be made in class or purchased. For information about non-toxic pest control, see the Mother Earth News web page, Organic Insect Control for Your Container Garden, at *www.motherearthnews.com/ organic-gardening/organic-insect-control-for-your-container-vegetable-garden*.

Observation Journal

Students create an observation journal in this lesson. An observation journal is a place for students to record data about their garden over time. In this lesson, students work collaboratively to decide what observations they should make and to create the journal. For information about plant observation journals and to access an online forum for sharing observations, visit the American Association for the Advancement of Science web page, The Science of Spring: Plant, Grow, Learn, at *http://seeds.sciencenetlinks.com*.

COMMON MISCONCEPTIONS

Students will have various types of prior knowledge about the concepts introduced in this lesson. Table 4.9 (p. 90) outlines some common misconceptions students may have concerning these concepts. Because of the breadth of students' experiences, it is not possible to anticipate every misconception that students may bring as they approach this lesson. Incorrect or inaccurate prior understanding of concepts can influence student learning in the future, however, so it is important to be alert to misconceptions such as those presented in the table.

Table 4.9. Common Misconceptions About the Concepts in Lesson 3

Topic	Student Misconception	Explanation
Engineers and the engineering design process (EDP)	Engineers' most important job is to build things.	While engineers do build machines, products, and structures, they must first go through a careful process of researching and planning before they begin to build. By doing this, they can be sure that the products they create meet people's needs in the best way possible. In fact, after engineers have completed building a machine or product, they will often test it and go back to the Learn and Plan stages of the EDP to improve their designs.

PREPARATION FOR LESSON 3

Review the Teacher Background Information provided, assemble the materials for the lesson, and preview the videos recommended in the Learning Components section below. If your garden will be located outdoors, check the weather report and plan accordingly.

Students will record their predictions for how their plants will grow in STEM Research Notebook entry #17. Two versions of this entry are provided in Appendix A; one for use if students plant seeds (#17a) and one for use if students plant seedlings (#17b).

Students will create a schedule for maintaining their garden using STEM Research Notebook entry #19. This schedule should be created on a weekly basis during the garden's growing season so that teams can rotate between various tasks.

In preparation for creating a class garden observation journal, students should observe plants on the school grounds. Each team should observe one type of plant. In addition to making preparations for taking students outdoors, you should become familiar with the types of plants on the school grounds so that you can assign each team a plant to observe.

LEARNING COMPONENTS
Introductory Activity/Engagement

Connection to the Challenge: Begin each day of this lesson by directing students' attention to the module challenge, the Container Garden Design Challenge:

> *Your town's leaders have noticed that many of the fruits, vegetables, and flowers sold in your grocery stores were grown far away from your local community. They*

think that growing more of these items locally will provide produce that is fresher and less expensive. Because of this, the town's leaders are looking for ways for families, schools, and businesses to use space creatively to grow fruits, vegetables, and flowers for members of the community to use. Your class has been challenged to create a garden to grow produce or other plants at your school as a model for the community of how this can be done and how your local weather patterns affect plants.

Hold a brief class discussion of how students' learning in the previous days' lessons contributed to their ability to complete the challenge. You may wish to create a class list of key ideas on chart paper.

Mathematics and Science Classes and ELA Connection: Remind students of their progress through the Define, Learn, and Plan phases of the EDP, asking them to share what they did during each of these phases. Tell students that in this lesson, they will complete the Try phase of the EDP as they build and plant their container garden.

Ask students to share what they know and what they want to know about how seeds turn into plants, documenting student responses on a KWL chart. Conduct an interactive read-aloud of *From Seed to Plant,* by Gail Gibbons, through which students will explore the progression from seed to plant.

STEM Research Notebook Entry #16

After the read-aloud, ask students to document what they learned about seeds and plants in their STEM Research Notebooks, using both words and pictures.

Before students begin to build and plant the container garden, ask them the following questions, depending on whether they are planting seeds or seedlings in their garden. Document student responses on chart paper and also have them write their responses as the next STEM Research Notebook entry:

- If they are planting seeds in their garden, ask students, What do your seeds need to sprout into seedlings? (Answers should be light and water.) How long do you think it will take for your seeds to sprout into seedlings? How much do you think your seedlings will grow each week?

- If they are planting seedlings, ask students, What do your seedlings need to grow? How long do you think it will take for your plants to grow 1 inch? How much do you think your seedlings will grow each week in your container garden?

STEM Research Notebook Entry #17

Have students record their responses to the questions about what their seeds or seedlings need to grow and how much and how quickly they will grow.

Social Studies Connection: Not applicable.

Activity/Exploration

Mathematics and Science Classes: Tell students that now they will use the plans the class made in Lesson 2 to build their container garden. Students should rely on the sketches and decisions they made as a class as they build the garden, using the appropriate materials from the list, depending on their design for the container garden, and taking measurements as necessary with measuring tapes or yardsticks. Prompt students to incorporate drainage holes in the bottom of the container so that excess water does not accumulate, which can cause plant diseases such as root rot. Then, students should fill the container with potting soil with their hands, while wearing protective gloves, and plant the seeds or seedlings.

Ask students how they will keep track of what is happening in their garden over time. Introduce the idea that scientists and other investigators collect data, or information, about things they investigate. Explain to students that they will complete the last phases of the EDP, Test and Decide, as they care for and observe their garden throughout the remainder of the school year.

Tell students that to record their ongoing observations of their garden, they will create a class observation journal. To prepare for this, students will make observations of plants they observe on the school grounds, and then use these observations to consider what data they should collect about their garden. Assign each team a particular plant to observe on the school grounds and provide each student with a clipboard and a ruler to use when recording their observations in their STEM Research Notebooks. Students should use crayons or colored pencils to add color to their drawings.

STEM Research Notebook Entry #18

Have students record their observations of their team's assigned plant, looking for evidence that its basic needs are being met and providing information about the plant's size and how the plant is responding to the season.

After completing their observations in the schoolyard, each team should pair with one other team to share observations. Next, the class should collaboratively decide what observations to make about their garden, using their experiences while observing plants on the school grounds as a guide. Students should consider how to organize these observations to track how plants develop throughout the year. Guide students to create charts or data sheets on which to record their plant data. For example, observations may include the following:

- Type of plant

- Height

- Time of observation

- Growth as related to the number of sunny days

- Growth as related to temperature

- Amount and type of pest control added

- Insect or animal damage

- Frequency of watering

- Amount of produce harvested

To construct an evidence-based account that young plants are like their parent plants, but not exactly like them, the observation journal can also include information about features that young and older plants share and sketches of various-sized leaves and other plant parts, such as buds and flowers.

In addition, the observation journal can include a section for the garden design. Remind students that the EDP is an ongoing process and that the Test phase for their design will include how well their plants grow and how easily their garden can be maintained. The first weeks of the garden's growth after it has been planted can be considered the Test phase of the EDP. Students collect data on plant growth and health during this time, and after students have collected about two weeks' worth of data on their garden, the class can analyze the data, calculating how much each plant has grown and how healthy each plant is (whether its basic needs are being met). As part of the Test phase, ask students to respond to the following questions:

- How much did our _____ plants grow? (ask for each type of plant)

- Are our plants healthy looking? Do you think their basic needs are being met?

- Are some plants growing better than others? Why do you think this is?

- How is our watering system working? (if the class incorporated this element)

An additional option for the Test phase is to invite a representative from a local garden center, farmer, or horticulturist to give the class feedback on the garden's design and plant growth.

After conducting observations for about a month, students should compare their garden's growth with the predictions they made in STEM Research Notebook Entry #17.

Students should also consider ways that the garden design might be modified to improve plant health and facilitate maintenance of the garden (Decide phase of the EDP). The class should use their findings from the Test phase for the Decide phase. Ask students if, based on their observations and their answers to the questions above, they think they could improve anything about the garden. What could they change about their garden to help the plants grow better? Based on their answers, the class can come up with a plan to make changes to the garden or the watering system to improve plant growth or health. The class should continue collecting data to see how the plants grow after students have made the improvements.

ELA Connection: Conduct an interactive read-aloud of *How a Seed Grows,* by Helene J. Jordan, which will allow students to continue exploring how seeds grow. Then, hold a class discussion about what students learned about seed growth, asking them to compare the seeds and plants highlighted in the book with the plants they observed on the school grounds.

Social Studies Connection: Not applicable.

Explanation

Mathematics and Science Classes: You may wish to have students transplant their lima bean and sunflower plants into their garden at this time. Then, explain to students that gardens need regular maintenance. This includes regularly watering the seeds or plants, applying fertilizer, protecting the plants from insects using nontoxic homemade insect repellents, and replacing any plants that did not survive. To continue to care for their plants on an ongoing basis after this module ends, students should create a schedule outlining which team will do the various tasks each day.

STEM Research Notebook Entry #19

Have students create a schedule to maintain the container garden, rotating student teams for various tasks, including watering, pest control, harvesting, and caring for tools.

Lesson Assessment. Assess student learning for the module by having students do the following:

- Identify two things seeds need to sprout into seedlings.

- Identify two parts of a plant that change as it grows.

- Identify two benefits of container gardening.

ELA Connection: Not applicable.

Social Studies Connection: Not applicable.

Elaboration/Application of Knowledge

Mathematics and Science Classes and ELA and Social Studies Connections: The observation journal can be used as part of the extension of the project into the remainder of the school year. Students should continue to maintain the observation journal as their plants grow. On a regular ongoing basis, they should graph or chart garden data, including such things as the height of plants, amount of produce harvested, growth as related to the number of sunny days and temperature, and any other data they decided to record in the journal.

In addition, students should continue to work through the last two steps of the EDP, Test and Decide, as they observe and maintain their garden. In particular, they should connect garden design with plant health and maintenance, making decisions about how the design could be modified to improve plant health or facilitate maintenance of the garden. Students should include notes about the ongoing EDP in the observation journal.

Once the garden has reached maturity, you may wish to have students share the class's garden design. This can be done in several different ways. For example, you could have students give a tour of the garden to other classes, parents, or community members, describing features of the garden design, the types of plants included, and how the class ensured that the plants' basic needs were met. Another option is for the class to create a blog or newsletter that describes students' progress in creating, maintaining, and improving the garden and provides advice to other community members who may wish to construct a container garden.

As students continue to observe and record data about their plants on an ongoing basis, show students a time lapse video of the life cycle of a plant, such as "Life Cycle Bean Plant 80" at *www.youtube.com/watch?v=fInh3iWv1LE*. Then, conduct an interactive read-aloud of *Plants!* by Brenda Iasevoli.

STEM Research Notebook Entries #20–23

Have students record their responses in these entries before and after they view the video and participate in the read-aloud.

A list of book suggestions for additional reading is included at the end of this chapter.

Evaluation/Assessment

Students may be assessed on the following performance tasks and other measures listed.

Performance Tasks

- Completion of Container Garden Design Challenge

- Observation journal

- Lesson Assessment

Other Measures (using assessment rubric in Appendix B, p. 134)

- Teacher observations
- STEM Research Notebook entries
- Participation in teams during investigations

INTERNET RESOURCES

Successful Container Gardens: Drainage web page
- *www.extension.illinois.edu/containergardening/choosing_drainage.cfm*

Successful Container Gardens: Fertilizers web page
- *www.extension.illinois.edu/containergardening/fertilizing.cfm*

Organic Insect Control for Your Container Garden web page
- *www.motherearthnews.com/organic-gardening/organic-insect-control-for-your-container-vegetable-garden*

The Science of Spring: Plant, Grow, Learn web page
- *http://seeds.sciencenetlinks.com*

"Life Cycle Bean Plant 80" video
- *www.youtube.com/watch?v=fInh3iWv1LE*

SUGGESTED BOOKS

- *The Science Book of Things That Grow,* by Neil Ardley (Harcourt Brace, 1991)
- *A Seed Is Sleepy,* by Dianna Hutts Aston (Chronicle Books, 2014)
- *Sprouting Seed Science Projects,* by Ann Benbow and Colin Mably (Enslow Elementary, 2009)
- *Leaves,* by Vijaya Khisty Bodach (Capstone Press, 2016)
- *The Tiny Seed,* by Eric Carle (Little Simon, 2009)
- *Flowers,* by John Farndon (Blackbirch Press, 2006)
- *From Seed to Plant,* by Allan Fowler (Children's Press, 2001)
- *The Reasons for Seasons,* by Gail Gibbons (Holiday House, 1995)

- *The Vegetables We Eat,* by Gail Gibbons (Holiday House, 2008)

- *The Reason for a Flower,* by Ruth Heller (Grosset & Dunlap, 1992)

- *The Carrot Seed,* by Ruth Krauss (HarperCollins, 2004)

- *Why Do Leaves Change Color?* by Betsy Maestro (HarperCollins, 2015)

- *A Fruit Is a Suitcase for Seeds,* by Jean Richards (Millbrook Press, 2002)

- *Seasons,* by Paul P. Sipiera and Diane M. Sipiera (Children's Press, 1999)

REFERENCE

Koehler, C., M. A. Bloom, and A. R. Milner. 2015. The STEM Road Map for grades K–2. In *STEM Road Map: A framework for integrated STEM education,* ed. C. C. Johnson, E. E. Peters-Burton, and T. J. Moore, 41–67. New York: Routledge. *www.routledge.com/products/9781138804234.*

TRANSFORMING LEARNING WITH PATTERNS AND THE PLANT WORLD AND THE *STEM ROAD MAP CURRICULUM SERIES*

Carla C. Johnson

This chapter serves as a conclusion to the Patterns and the Plant World integrated STEM curriculum module, but it is just the beginning of the transformation of your classroom that is possible through use of the *STEM Road Map Curriculum Series.* In this book, many key resources have been provided to make learning meaningful for your students through integration of science, technology, engineering, and mathematics, as well as social studies and English language arts, into powerful problem- and project-based instruction. First, the Patterns and the Plant World curriculum is grounded in the latest theory of learning for students in grade 1 specifically. Second, as your students work through this module, they engage in using the engineering design process (EDP) and build prototypes like engineers and STEM professionals in the real world. Third, students acquire important knowledge and skills grounded in national academic standards in mathematics, English language arts, science, and 21st century skills that will enable their learning to be deeper, retained longer, and applied throughout, illustrating the critical connections within and across disciplines. Finally, authentic formative assessments, including strategies for differentiation and addressing misconceptions, are embedded within the curriculum activities.

The Patterns and the Plant World curriculum in The Represented World STEM Road Map theme can be used in single-content classrooms (e.g., mathematics) where there is only one teacher or expanded to include multiple teachers and content areas across classrooms. Through the exploration of the Container Garden Design Challenge, students

engage in a real-world STEM problem on the first day of instruction and gather necessary knowledge and skills along the way in the context of solving the problem.

The other topics in the *STEM Road Map Curriculum Series* are designed in a similar manner, and NSTA Press has additional volumes in this series for this and other grade levels and plans to publish more. The volumes covering Innovation and Progress have been published and are as follows:

- *Amusement Park of the Future, Grade 6*

- *Construction Materials, Grade 11*

- *Harnessing Solar Energy, Grade 4*

- *Transportation in the Future, Grade 3*

- *Wind Energy, Grade 5*

The tentative list of other books includes the following themes and subjects:

- The Represented World

  - Car crashes

  - Changes over time

  - Improving bridge design

  - Packaging design

  - Radioactivity

  - Rainwater analysis

  - Swing set makeover

- Cause and Effect

  - Influence of waves

  - Hazards and the changing environment

  - The role of physics in motion

- Sustainable Systems

  - Creating global bonds

  - Composting: Reduce, reuse, recycle

  - Hydropower efficiency

  - System interactions

- Optimizing the Human Experience

  - Genetically modified organisms

  - Mineral resources

  - Rebuilding the natural environment

  - Water conservation: Think global, act local

If you are interested in professional development opportunities focused on the STEM Road Map specifically or integrated STEM or STEM programs and schools overall, contact the lead editor of this project, Dr. Carla C. Johnson (*carlacjohnson@purdue.edu*), associate dean and professor of science education at Purdue University. Someone from the team will be in touch to design a program that will meet your individual, school, or district needs.

APPENDIX A

STEM RESEARCH NOTEBOOK TEMPLATES

MY STEM RESEARCH NOTEBOOK

OUR CONTAINER GARDEN

Name:

_____

- -

_____

Name: _____   Date: _____

STEM RESEARCH NOTEBOOK ENTRY #1 (LESSON PLAN 1)

I learned ...

Name: _____   Date: _____

STEM RESEARCH NOTEBOOK ENTRY #2 (LESSON PLAN 1)

I learned ...

NATIONAL SCIENCE TEACHERS ASSOCIATION

Name: _____   Date: _____

STEM RESEARCH NOTEBOOK ENTRY #3 (LESSON PLAN 1)

I learned ...

_____

- -

_____

_____

- -

_____

_____

- -

_____

Name: _____   Date: _____

STEM RESEARCH NOTEBOOK ENTRY #4, PAGE 1 (LESSON PLAN 1)

TERRIFIC SEASONAL TREE

What is the weather usually like in this season in your habitat?

Name: _____   Date: _____

STEM RESEARCH NOTEBOOK ENTRY #4, PAGE 2 (LESSON PLAN 1)

TERRIFIC SEASONAL TREE

How does this season affect the basic needs of the living things in your habitat? (Consider air, water, food, habitat, and sunlight.)

What happens to your tree during this season?

Name: _____

Date: _____

STEM RESEARCH NOTEBOOK ENTRY #5 (LESSON PLAN 1)

VOCABULARY WORDS

Key Vocabulary Word	Definition	Illustration

Name: _____   Date: _____

STEM RESEARCH NOTEBOOK ENTRY #6 (LESSON PLAN 1)

I learned ...

Name: _____   Date: _____

STEM RESEARCH NOTEBOOK ENTRY #7 (LESSON PLAN 2)

I learned ...

Name: _____   Date: _____

STEM RESEARCH NOTEBOOK ENTRY #8 (LESSON PLAN 2)

I learned ...

_____

- -

_____

_____

- -

_____

- -

_____

Name: _____   Date: _____

STEM RESEARCH NOTEBOOK ENTRY #9 (LESSON PLAN 2)

I learned ...

Name: _____   Date: _____

STEM RESEARCH NOTEBOOK ENTRY #10, PAGE 1 (LESSON PLAN 2)

PLAN: DRAW A DIAGRAM OF YOUR CONTAINER GARDEN DESIGN.

1. Draw the design of your container garden.

   - Decide what types of materials you will use to make your garden (e.g., wood, recycled milk jugs, recycled bottles or cans), and label those on your design.

Name: _____   Date: _____

STEM RESEARCH NOTEBOOK ENTRY #10, PAGE 2 (LESSON PLAN 2)

2. How will you water your garden? (Think about using recycled plastic bottles.) Draw your plan for watering and label the parts of the drawing.

3. How will you keep insects and pests away from your garden?

Name: _____   Date: _____

STEM RESEARCH NOTEBOOK ENTRY #11 (LESSON PLAN 2)

I learned ...

Name: _____ Date: _____

STEM RESEARCH NOTEBOOK ENTRY #12 (LESSON PLAN 2)

Investigate and identify supply needs for your container garden.

Questions	Answers
What supplies do we need to make our garden? (Think about the design you drew earlier. Did you use wood, recycled milk jugs, recycled bottles, or recycled cans? What about art supplies, soil, and plants?)	
How much soil do we need?	
What seeds or seedlings do we need?	
How many seeds or seedlings do we need?	
What tools do we need?	
How can we provide enough water? (Think about the design for a watering system that you drew in STEM Research Notebook Entry #10.)	

NATIONAL SCIENCE TEACHERS ASSOCIATION

Name: _____   Date: _____

STEM RESEARCH NOTEBOOK ENTRY #13 (LESSON PLAN 2)

Decide what supplies you can get for your container garden. How can you help provide supplies for the garden?

Items	How I Can Help
Soil	
Seeds or seedlings	
Tools	
Other supplies	
Watering system	

Name: _____   Date: _____

STEM RESEARCH NOTEBOOK ENTRY #14 (LESSON PLAN 2)

I learned ...

_____

- -

_____

_____

- -

_____

_____

- -

_____

Name: _____   Date: _____

STEM RESEARCH NOTEBOOK ENTRY #15 (LESSON PLAN 2)

What recycled items from school or items from home could you use to meet your garden's supply needs?

Materials From Supply List	Recycled Items or Items Brought From Home

Name: _____   Date: _____

_____

STEM RESEARCH NOTEBOOK ENTRY #16 (LESSON PLAN 3)

_____

I learned …

_____

- -

_____

_____

- -

_____

_____

- -

_____

Name: _____   Date: _____

STEM RESEARCH NOTEBOOK ENTRY #17A (LESSON PLAN 3)

PLANT LIFE CYCLE FOR SEEDS

Circle or write in your predictions.

Questions	Predictions
What do your seeds need to sprout into seedlings?	Nothing Light Water
How long do you think it will take for your seeds to sprout into seedlings?	About 1 week About 1 month About 1 year
How much do you think your seedlings will grow each week?	About 1 inch About 1 foot About 1 yard

Name: _____ Date: _____

STEM RESEARCH NOTEBOOK ENTRY #17B (LESSON PLAN 3)

PLANT LIFE CYCLE FOR SEEDLINGS

Circle or write in your predictions.

Questions	Predictions
What do your seedlings need to grow?	Nothing Light Water
How long do you think it will take for your seedlings to grow 1 inch?	About 1 week About 1 month About 1 year
How much do you think your seedlings will grow each week?	About 1 inch About 1 foot About 1 yard

Name: _____   Date: _____

STEM RESEARCH NOTEBOOK ENTRY #18, PAGE 1 (LESSON PLAN 3)

Complete the following using words and pictures with color:

1. My team observed this plant (draw a picture and write the plant's name):

2. Draw a picture of one of this plant's leaves:

3. Draw a picture of this plant's stem:

4. Does the plant look as if all its basic needs are being met? (A plant's basic needs are air, water, food, habitat, and sunlight.) How can you tell?

Name: _____   Date: _____

STEM RESEARCH NOTEBOOK ENTRY #18, PAGE 2 (LESSON PLAN 3)

5. What about the way this plant looks gives you a clue about what season it is?

6. Can you measure this plant with a ruler? If so, how tall is it in inches? If you cannot measure it with a ruler, compare its size with something close by, and draw a picture with labels showing its size:

Name: _____   Date: _____

STEM RESEARCH NOTEBOOK ENTRY #19 (LESSON PLAN 3)

Maintaining and sustaining the garden (rotate various student teams each week)

Daily schedule:

- Watering
- Pest control
- Harvesting
- Wash tools and put away after use
- Other tasks the class agrees on

Task	Team Responsible for Task				
	Monday	**Tuesday**	**Wednesday**	**Thursday**	**Friday**
Watering					
Pest Control					
Harvesting					
Wash tools and put away					
Other:					

Name: _____   Date: _____

STEM RESEARCH NOTEBOOK ENTRY #20 (LESSON PLAN 3)

Respond to the following *before* viewing the video:

I want to know …

_____

_____

_____

_____

_____

Name: _____   Date: _____

STEM RESEARCH NOTEBOOK ENTRY #21 (LESSON PLAN 3)

Respond to the following *after* viewing the video:

I learned ...

Name: _____   Date: _____

STEM RESEARCH NOTEBOOK ENTRY #22 (LESSON PLAN 3)

Respond to the following *before* the read-aloud:

I want to know ...

_____

- -

_____

- -

_____

- -

_____

- -

_____

Name: _____   Date: _____

STEM RESEARCH NOTEBOOK ENTRY #23 (LESSON PLAN 3)

Respond to the following *after* the read-aloud:

I learned ...

_____

- -

_____

_____

- -

_____

_____

- -

_____

APPENDIX B

OBSERVATION, STEM RESEARCH NOTEBOOK, AND PARTICIPATION RUBRIC

Observation, STEM Research Notebook, and Participation Rubric

Name: _____

Categories (components)	Missing or Unrelated (0 points)	Beginning (1 point)	Developing (2 points)	Meets Expectations (3 points)	Exceeds Expectations (4 points)	Score
OBSERVATION OF LISTENING AND DISCUSSION SKILLS	Component is missing or unrelated.	Does not listen to others and shows little respect for alternative viewpoints.	Occasionally listens to others but often speaks out of turn.	Listens to others, only occasionally speaks out of turn, and generally accepts other points of view.	Listens carefully to others, waits for turn to speak, and respects alternative viewpoints.	
STEM RESEARCH NOTEBOOK	Component is missing or unrelated.	Indicates little understanding of the concepts being taught.	Recalls and is able to explain basic facts and concepts.	Demonstrates ability to apply concepts, using information in new situations.	Demonstrates a deep understanding of concepts by drawing relationships between ideas and using information to generate new ideas.	
PARTICIPATION	Component is missing.	Does not volunteer. When responding to teacher prompts, comments are sometimes not relevant to the discussion.	Responds to teacher prompts during classroom discussions but does not volunteer.	Willingly participates in classroom discussions and offers relevant comments.	Contributes insightful comments and poses thoughtful questions.	

TOTAL SCORE: _____

COMMENTS:

APPENDIX C

CONTENT STANDARDS ADDRESSED IN THIS MODULE

NEXT GENERATION SCIENCE STANDARDS

Table C1 (p. 136) lists the science and engineering practices, disciplinary core ideas, and crosscutting concepts this module addresses. The supported performance expectations are as follows:

- 1-ESS1-1. Use observations of the sun, moon, and stars to describe patterns that can be predicted.

- 1-ESS1-2. Make observations at different times of year to relate the amount of daylight to the time of year.

- 1-LS1-2. Read texts and use media to determine patterns in behavior of parents and offspring that help offspring survive.

- 1-LS3-1. Make observations to construct an evidence-based account that young plants and animals are like, but not exactly like, their parents.

Table C1. *Next Generation Science Standards (NGSS)*

Science and Engineering Practices

PLANNING AND CARRYING OUT INVESTIGATIONS

Planning and carrying out investigations to answer questions or test solutions to problems in K–2 builds on prior experiences and progresses to simple investigations, based on fair tests, which provide data to support explanations or design solutions.

- Plan and conduct investigations collaboratively to produce evidence to answer a question.

ANALYZING AND INTERPRETING DATA

Analyzing data in K–2 builds on prior experiences and progresses to collecting, recording, and sharing observations.

- Use observations (firsthand or from media) to describe patterns in the natural world in order to answer scientific questions.

CONSTRUCTING EXPLANATIONS AND DESIGNING SOLUTIONS

Constructing explanations and designing solutions in K–2 builds on prior experiences and progresses to the use of evidence and ideas in constructing evidence-based accounts of natural phenomena and designing solutions.

- Make observations (firsthand or from media) to construct an evidence-based account for natural phenomena.

OBTAINING, EVALUATING, AND COMMUNICATING INFORMATION

Obtaining, evaluating, and communicating information in K–2 builds on prior experiences and uses observations and texts to communicate new information.

- Read grade-appropriate texts and use media to obtain scientific information to determine patterns in the natural world.

Disciplinary Core Ideas

ESS1.A: THE UNIVERSE AND ITS STARS

- Patterns of the motion of the sun, moon, and stars in the sky can be observed, described, and predicted.

ESS1.B: EARTH AND THE SOLAR SYSTEM

- Seasonal patterns of sunrise and sunset can be observed, described, and predicted.

LS1.A: STRUCTURE AND FUNCTION

- All organisms have external parts. Different animals use their body parts in different ways to see, hear, grasp objects, protect themselves, move from place to place, and seek, find, and take in food, water and air. Plants also have different parts (roots, stems, leaves, flowers, fruits) that help them survive and grow.

Continued

Table C1. (*continued*)

LS1.B: GROWTH AND DEVELOPMENT OF ORGANISMS • Adult plants and animals can have young. In many kinds of animals, parents and the offspring themselves engage in behaviors that help the offspring to survive. **LS1.D: INFORMATION PROCESSING** • Animals have body parts that capture and convey different kinds of information needed for growth and survival. Animals respond to these inputs with behaviors that help them survive. Plants also respond to some external inputs. **LS3.A: INHERITANCE OF TRAITS** • Young animals are very much, but not exactly, like their parents. Plants also are very much, but not exactly, like their parents. **LS3.B: VARIATION OF TRAITS** • Individuals of the same kind of plant or animal are recognizable as similar but can also vary in many ways.
Crosscutting Concepts
PATTERNS • Patterns in the natural world can be observed, used to describe phenomena, and used as evidence. **STRUCTURE AND FUNCTION** • The shape and stability of structures of natural and designed objects are related to their function(s).

Source: NGSS Lead States. 2013. *Next Generation Science Standards: For states, by states.* Washington, DC: National Academies Press. *www.nextgenscience.org/next-generation-science-standards.*

Table C2. Common Core Mathematics and English Language Arts (ELA) Standards

MATHEMATICAL PRACTICES

- MP1. Make sense of problems and persevere in solving them.
- MP2. Reason abstractly and quantitatively.
- MP3. Construct viable arguments and critique the reasoning of others.
- MP4. Model with mathematics.
- MP5. Use appropriate tools strategically.
- MP6. Attend to precision.
- MP7. Look for and make use of structure.
- MP8. Look for and express regularity in repeated reasoning.

MATHEMATICAL CONTENT

- 1.NBT.B.3. Compare two two-digit numbers based on meanings of the tens and ones digits, recording the results of comparisons with the symbols >, =, and <.
- 1.MD.A.1. Order three objects by length; compare the lengths of two objects indirectly by using a third object.
- 1.MD.C.4. Organize, represent, and interpret data with up to three categories; ask and answer questions about the total number of data points, how many in each category, and how many more or less are in one category than in another.
- 1.OA.A.1. Use addition and subtraction within 20 to solve word problems involving situations of adding to, taking from, putting together, taking apart, and comparing, with unknowns in all positions.
- 1.OA.A.2. Solve word problems that call for addition of three whole numbers whose sum is less than or equal to 20, e.g., by using objects, drawings, and equations with a symbol for the unknown number to represent the problem.

READING STANDARDS

- RI.1.1. Ask and answer questions about key details in a text.
- RI.1.2. Identify the main topic and retell key details of a text.
- RI.1.3. Describe the connection between two individuals, events, ideas, or pieces of information in a text.
- RI.1.7. Use the illustrations and details in a text to describe its key ideas.

WRITING STANDARDS

- W.1.2. Write informative/explanatory texts in which they name a topic, supply some facts about the topic, and provide some sense of closure.
- W.1.6. With guidance and support from adults, use a variety of digital tools to produce and publish writing, including in collaboration with peers.
- W.1.7. Participate in shared research and writing.
- W.1.8. With guidance and support from adults, recall information from experiences or gather information from provided sources to answer a question.

SPEAKING AND LISTENING STANDARDS

- SL.1.1. Participate in collaborative conversations with diverse partners about *grade 1 topics and texts* with peers and adults in small and larger groups.
- SL.1.1.A. Follow agreed-upon rules for discussions.
- SL.1.1.B. Build on others' talk in conversations by responding to the comments of others through multiple exchanges.
- SL.1.1.C. Ask questions to clear up any confusion about the topics and texts under discussion.
- SL.1.3. Ask and answer questions about what a speaker says in order to gather additional information or clarify something that is not understood.
- SL.1.5. Add drawings or other visual displays to descriptions when appropriate to clarify ideas, thoughts, and feelings.

Source: National Governors Association Center for Best Practices and Council of Chief State School Officers (NGAC and CCSSO). 2010. *Common core state standards.* Washington, DC: NGAC and CCSSO.

Table C3. National Association for the Education of Young Children (NAEYC) Standards

NAEYC Curriculum Content Area for Cognitive Development: Science and Technology
• 2.E.1. Arrange firsthand, meaningful experiences that are intellectually and creatively stimulating, invite exploration and investigation, and engage children's active, sustained involvement by providing a rich variety of material, challenges, and ideas.
• 2.F.3. Extend the range of children's interests and the scope of their thought, present novel experiences and introduce stimulating ideas, problems, experiences, or hypotheses.
• 2.F.6. Enhance children's conceptual understanding through various strategies, including intensive interview and conversation, encourage children to reflect on and "revisit" their experiences.
• 2.G.2. Scaffolding takes on a variety of forms.
• 2.J.1. Incorporate a wide variety of experiences, materials and equipment, and teaching strategies to accommodate the range of children's individual differences in development, skills and abilities, prior experiences, needs, and interests.
• 3.A.1. Teachers consider what children should know, understand, and be able to do across the domains.

Source: National Association for the Education of Young Children (NAEYC). 2005. *NAEYC early childhood program standards and accreditation criteria: The mark of quality in early childhood education.* Washington, DC: NAEYC.

Table C4. 21st Century Skills From the Framework for 21st Century Learning

21st Century Skills	Learning Skills and Technology Tools	Teaching Strategies	Evidence of Success
INTERDISCIPLINARY THEMES	• Economic, Business, and Entrepreneurial Literacy • Health Literacy • Environmental Literacy	• Provide students with the opportunity to investigate the life cycles of plants in the context of the business, economics, and industry of everyday life (e.g., gardening, farming, agriculture).	• Students communicate their prior experiences with the life cycles of plants in the context of everyday life.
LEARNING AND INNOVATION SKILLS	• Creativity and Innovation • Critical Thinking and Problem Solving • Communication and Collaboration	• Facilitate creativity and innovation through having students design and create a container garden and observation journal. • Facilitate critical thinking and problem solving through use of the EDP and having students make observations in a real-world setting about changing plant life.	• Students demonstrate creativity and innovation, critical thinking, and problem solving as they develop a container garden and observation journal. • Students work collaboratively and communicate effectively in teams to complete a group project.
INFORMATION, MEDIA, AND TECHNOLOGY SKILLS	• Information Literacy • Media Literacy • ICT Literacy	• Engage students in guided practice and scaffolding strategies through the use of developmentally appropriate books, videos, and websites to advance their knowledge.	• Students acquire and use deeper content knowledge as they work to complete the project and keep an observation journal.
LIFE AND CAREER SKILLS	• Flexibility and Adaptability • Initiative and Self-Direction • Social and Cross-Cultural Skills • Productivity and Accountability • Leadership and Responsibility	• Facilitate student collaborative teamwork to foster life and career skills.	• Throughout this module, students collaborate to conduct research and work on their group project.

Source: Partnership for 21st Century Learning. 2015. Framework for 21st Century Learning. *www.p21.org/our-work/p21-framework.*

Table C5. English Language Development (ELD) Standards

ELD STANDARD 1: SOCIAL AND INSTRUCTIONAL LANGUAGE

English language learners communicate for Social and Instructional purposes within the school setting.

ELD STANDARD 2: THE LANGUAGE OF LANGUAGE ARTS

English language learners communicate information, ideas, and concepts necessary for academic success in the content area of Language Arts.

ELD STANDARD 3: THE LANGUAGE OF MATHEMATICS

English language learners communicate information, ideas, and concepts necessary for academic success in the content area of Mathematics.

ELD STANDARD 4: THE LANGUAGE OF SCIENCE

English language learners communicate information, ideas, and concepts necessary for academic success in the content area of Science.

ELD STANDARD 5: THE LANGUAGE OF SOCIAL STUDIES

English language learners communicate information, ideas, and concepts necessary for academic success in the content area of Social Studies.

Source: WIDA. 2012. 2012 amplification of the English language development standards: Kindergarten–grade 12. *https://wida.wisc.edu/teach/standards/eld.*

INDEX

Page numbers printed in **boldface type** indicate tables, figures, or handouts.

NATIONAL SCIENCE TEACHERS ASSOCIATION